PRAISE FOR

Cosmogene[illegible]

"*Cosmogenesis* is one of the most remarkable s[illegible] a page-turner, but a life-changer. We follow Swimme's journey of studying science as an academic field to embracing cosmology as a lived experience. In reading this, you will be transformed as you enter into new ways of becoming alive in the universe."

—MARY EVELYN TUCKER, senior lecturer at Yale University and translator and editor of *The Philosophy of Qi*

"In *Cosmogenesis*, Brian Thomas Swimme, one of the preeminent Western thinkers on cosmology and science, explores the history of the universe, connecting the dots in mythology, philosophy, and science. In a sweeping arc from the origin of the universe to the challenges of the present day, he portrays us for what we really are: cosmological beings and a tribe of storytellers. He shows that storytelling itself is a cosmic event, a moment of cosmic self-awareness. This is a book in the tradition of grand narrative, and a must-read for the curious, searching for our place in the universe and our mission for the cosmic future."

—WOLFGANG LEIDHOLD, professor of political theory and history of ideas at University of Cologne

"*Cosmogenesis* has the gripping intensity of a mystery thriller and the Socratic seriousness of a truth quest. In the pages of this book, the staggering epic of a fourteen-billion-year evolution is told via a steadily accelerating drumbeat of revelations. Both remarkably accessible and irresistibly seductive, this transformative work augurs spiritual rebirth at a planetary scale."

—LOUIS G. HERMAN, author of *Future Primal*

"*Cosmogenesis* is the most clearly and convincingly told story of the origin of the universe I have ever read. But that's not all that makes this an essential read. If humanity is to escape the stranglehold of the lonely, king-of-the-mountain materialism that licenses planetary destruction, it will be when we tell a new story about who we are and how we come to be in the universe expanding around and within us. Cosmogenesis is a powerful telling of that story."

—KATHLEEN DEAN MOORE, author of *Great Tide Rising*

"*Cosmogenesis* is not only an engaging memoir of a visionary cosmologist, but an autobiographical coming-of-age story of the cosmos itself."

—SEAN KELLY, PHD, author of *Becoming Gaia: On the Threshold of Planetary Initiation* and *Coming Home: The Birth & Transformation of the Planetary Era*

"A love story beautifully told about the storyteller and his vocation, his wife, his mentor Thomas Berry, the universe, and us. Wonderful snapshot moments enliven the soul and imagination and tell the valuable and vulnerable story of the price paid for coming to one's vocation. At the same time, the storyteller heralds a new direction for humanity, freshly empowered by our role as cosmic beings in a sacred process of cosmogenesis, where our vocations, too, become love stories born of our sacred origins."

—MATTHEW FOX, author of *Original Blessing, The Coming of the Cosmic Christ*, and *The Tao of Thomas Aquinas*

"An immediate antidote to shrunken consciousness, *Cosmogenesis* will enlarge your conception of what a human is for. Swimme's riveting auto-cosmology offers a direct, repeatable experience of intimacy with the universe."

—CAROLYN COOKE, author of *Daughters of the Revolution* and *Amor and Psycho*

"As a theoretical physicist interested in the wider cultural and spiritual implications of modern science, I have long been a fan of the work of Brian Thomas Swimme, who has a particularly inspiring take on the standard scientific paradigm—that the universe is not just something to be theoretically understood but something we can have an experiential relationship with. In his newest book, *Cosmogenesis*, he intermingles his description of the development of his ideas with his own very personal history. Poignant, epic, and fascinating, it is quite simply a joy to read."

—PROFESSOR JONATHAN HALLIWELL, Department of Physics, Imperial College London

"*Cosmogenesis* is a remarkable story threaded into the dynamic unfolding of our emerging universe. The weave here of poetics, spirituality, and science is exquisite. Swimme's narration will carry you along in his poignant journey toward universe-as-teacher. You will discover your personal entry into a cosmos speaking through you."

—JOHN GRIM, senior lecturer at Yale University and coauthor of *Ecology and Religion*

"In *Cosmogenesis*, Brian Thomas Swimme revisits the personal stages of his journey of discovery from his beginning as an academic mathematician to his exceptional bursts of insight that comprise his truth quest as a cosmologist. For Swimme, these insights inaugurate a new multi-civilizational Axial Age whose recognition will come sooner than the original one twenty-five hundred years ago, since the explosive natural and political crises syndromes are global and do not allow for much delay."

—MANFRED HENNINGSEN, professor emeritus of political science at University of Hawaii at Manoa

"In Swimme's sweeping and beautiful memoir of a life in search of truth, the reader is witness to a remarkable feat of intellectual alchemy. As our protagonist reaches ever deeper for the genesis of the universe, as revealed in the mathematics of his native tongue, he uncovers a subtle, stunning truth. The universe 'out there' is intricately and intrinsically tied up with the universe 'in here.' Swimme's storytelling skill shines brightly in this unique intellectual journey, complementing the extraordinary words of, and the author's engaging encounters with, many of the great minds of the last century. *Cosmogenesis* is a monument to mystic chords of nature's memory whose notes join both consciousness and cosmos in the story of creation. May we all be lifted by the soaring, resonant voice of this cosmic bard and learn to sing in harmony with his hymn to the universe."

—CARTER PHIPPS, author of *Evolutionaries* and *Conscious Leadership*

"This is a brave, paradigm-shifting, and page-turning book; vulnerable and intimate, vast and lyrical. By sharing the challenges and breakthroughs of his own personal story in the style of an 'auto-cosmology,' Brian Thomas Swimme impactfully reveals epic twists and turns of the Universe Story. Powerful storytelling throughout evokes a new and much needed cosmology that meaningfully brings together subjective lived experience and recent scientific discovery."

—JANE RIDDIFORD, author of *Learning to Lead Together: An Ecological and Community Approach* and cofounder of Global Generation

"Modern science began with the myth of the machine. In only a few centuries, this worldview has completely transformed the world, empowering our species to become a geological force on par with asteroids and ice ages. If Brian Thomas Swimme and Thomas Berry are right, science, through the insights derived from its own methods, has now entirely outgrown the mechanistic image of the cosmos. The universe, it turns out, is a green dragon—a creative process rather than a finished product. Swimme's autobiocosmological narrative is, in one sense, the exemplar of an ancient genre counting Augustine's *Confessions* and Dante's *Divine Comedy* among its antecedents. But in another sense, *Cosmogenesis* changes everything: in place of a separate creator God who designs and judges creation from afar, there is pervasive creative attraction luring the universe beyond every settled order; and in place of a static hierarchy of planetary spheres, there is a nested sequence of evolutionary phases. The human is not the end of this stupendous process but a crucial turning point. Conscious, self-reflective creatures now hold the fate of Earth in their hands. Swimme's book serves as a call to adventure to all those ready and willing to meet this moment by becoming cosmological beings." —MATTHEW D. SEGALL, PHD, author of *Physics of the World-Soul*

"What a wonderfully engaging book Brian Thomas Swimme has given us! So spacious and expansive, even cosmic, yet down-to-earth and deeply human. The lightness of touch, the exquisite care for detail, the honesty, the self-deprecating humor: every scene is described with such Salinger-like vividness it's as if it is taking place now before one's eyes. At one level, *Cosmogenesis* is a model of how to communicate epoch-shaping scientific and philosophical discoveries while narrating the unfolding drama of one's own life story. But at a deeper level, it is the gradual revelation of how the life of the individual and the life of the cosmos are fundamentally intertwined within a single journey of self-revelation."

RICHARD TARNAS, author of *The Passion of the Western Mind* and *Cosmos and Psyche*

"In this book, without warning, a star-inspired mathematician summons us, squabbling bipeds of an overheating planet, to see ourselves as the evolving universe itself. The grandeur of that calling would seem ludicrously beyond our capacity were it not for the trust our guide elicits. Perhaps what moves us most is the wild generosity he helps us to see at work in the cosmos, and evident in the self-offering of the [super]nova that birthed our own solar system." —JOANNA MACY, author of *World as Lover, World as Self*

Cosmogenesis

ALSO BY BRIAN THOMAS SWIMME

The Universe Is a Green Dragon: A Cosmic Creation Story

Journey of the Universe with Mary Evelyn Tucker

The Universe Story: From the Primordial Flaring Forth to the Ecozoic Era: A Celebration of the Unfolding of the Cosmos with Thomas Berry

The Hidden Heart of the Cosmos: Humanity and the New Story

Manifesto for a Global Civilization with Matthew Fox

Cosmogenesis

An Unveiling of the Expanding Universe

Brian Thomas Swimme

Counterpoint
BERKELEY, CALIFORNIA

Cosmogenesis

First hardcover edition: 2022
First paperback edition: 2023

The Library of Congress has cataloged the hardcover edition as follows:
Names: Swimme, Brian, author.
Title: Cosmogenesis : an unveiling of the expanding universe / Brian Thomas Swimme.
Description: First hardcover edition. | Berkeley : Counterpoint, 2022.
Identifiers: LCCN 2021045936 | ISBN 9781640093980 (hardcover) | ISBN 9781640093997 (ebook)
Subjects: LCSH: Swimme, Brian. | Astronomers—United States—Biography. | Cosmology.
Classification: LCC QC858.S95 A3 2022 | DDC 520/.92 [B]—dc23/eng/20220201
LC record available at https://lccn.loc.gov/2021045936

Paperback ISBN: 978-1-64009-617-2

Cover design by Lexi Earle
Cover photograph © iStock / titoOnz
Book design by Laura Berry

COUNTERPOINT
2560 Ninth Street, Suite 318
Berkeley, CA 94710
www.counterpointpress.com

Printed in the United States of America
10 9 8 7 6 5 4 3 2 1

To Denise Marie Santi Swimme

Contents

PART TWO ✴ *Stars and Galaxies*

PART ONE

*

The Birth of the Universe

Prologue

Finding Our Bearings in the River of Time

In my life as a cosmologist, my mission has been to celebrate the great events of cosmogenesis by employing the central theories of contemporary science: quantum mechanics, the second law of thermodynamics, the general theory of relativity, plate tectonics, natural selection, encephalization. These theories have enabled us to discover our cosmic genesis, which can be summarized in a single complex sentence: the universe began fourteen billion years ago with the emergence of elementary particles in the form of primordial plasma, which quickly morphed into atoms of hydrogen, helium, and lithium; a hundred million years later, galaxies began to appear, and in one of these, the Milky Way, minerals arranged themselves into living cells that constructed advanced life, including evergreen trees, coral reefs, and the vertebrate nervous systems that humans used to discover this entire sequence of universe development. That sentence required four and a half centuries of scientific investigation of matter. Cosmogenesis is certainly among the ten most significant ideas in human history. It dwarfs Copernicus's discovery that the Sun is the center of the solar system.

After decades of teaching this evolutionary cosmology, I suddenly realized something. I had left myself out. The demands for getting the necessary knowledge to tell the stories of Earth and universe filled me so completely I forgot to include the story of the storyteller. To be honest, it was not just that I had forgotten, I had tricked myself into thinking cosmology was the story of how things "out there" evolved through time. But then I realized I was one of those things, that I was evolving, that I was as much a development of the universe as were stars and galaxies. If I wanted to tell the story of the expanding universe and how it developed through time, I needed to include the story of my long struggle out of the structures of existence I had been born into.

THIS BOOK lives in the overlap of two genres. It is in the tradition of autobiography in that it is a search for meaning that began explicitly in 1968 with Denise Santi and Bruce Bochte at Santa Clara University, that deepened with Dolores Maro at the University of Puget Sound and with Matthew Fox in Chicago, and that culminated in 1983 in New York City in a yearlong conversation with Thomas Berry. But in addition to being autobiographical, it is also in the tradition of cosmology in that these conversations were woven around the major features of the expanding universe.

Combining autobiography and cosmology has become a necessity. During the years before the discovery of cosmogenesis, we scientists could live in the fantasy of thinking we could step out of the universe in order to understand it. But there is no outside place where cosmologists can stand and see the nature of things. Every adequate cosmology must include the story of the narrator. To explore this idea, Professor Carolyn Cooke, my colleague at the California Institute of Integral Studies, and I constructed a series of courses around a neologism that brought the two genres together, *auto-cosmology*. Though Carolyn and I thought it was a new term, the word was used, and perhaps invented, in 1979 by Ursula K. Le Guin. Her term captures the new opportunity for those writers, artists, scientists, and musicians who have a cosmic sense, who see their work as involving the universe as a whole.

Before we begin, one last point needs to be made. Not one of the ten thousand previous generations of humans knew the cosmic sequence of transformations that brought us forth. Those of us exploring the meaning of this discovery are venturing onto untrodden land. It might take centuries for us to articulate an adequate orientation within the cosmic river in which we find ourselves. Which means that in addition to the thrill of voyaging into unmapped territory, we need to remember that any suggestion of having arrived at a final understanding is the presence of folly. My own story is offered with the hope that it might become a small part of an emerging, planetwide conversation on how to live together in this fourteen-billion-year enterprise.

1.

To Understand the Origin of All

I begin my story on August 23, 1978, my first day as a professional mathematician at the University of Puget Sound, in Tacoma, Washington. The night before, to be precise. Too excited for sleep, I lay in bed next to my wife, Denise, waiting out the hours, but then gave up and crept into the hallway. Due to the gravity of that moment, every tiny detail still lives in my memory. I can feel the glass knob of our bedroom door as I gently released the catch so it would not snap and disturb Denise's sleep. In the hallway I pushed the white lever of the thermostat to the right until I heard that *whomp* sound of our furnace kicking to life, then nudged it a bit farther to make sure the house would be plenty warm.

While the water heated on the stove, I stepped onto the front porch and looked for the stars and Moon, but they were obscured by the cloud cover blown in from the Pacific. The cedars, maples, chestnuts, sycamores, and elms that lined the streets had convinced us to buy. Tacoma had planted them in the 1800s with the arrival of Western civilization, and they had since grown into a crowning canopy. We were lucky to find a house two blocks east of the university. The former owner had also been a professor, in the Physics Department. At the signing he pointed to several nearby houses that had passed from professor to professor since their construction ninety years earlier. Across from us was Theodore Taranovski, from the Department of History, whom we would see late at night in his upstairs office, and farther north Dolores Maro, from the Classics Department.

When the kettle first began to bubble, I dashed to the kitchen to grab it before it screamed, and I poured the water over the coffee grounds, adding half-and-half until it turned tawny brown. Positioning myself at my grandmother's pinewood table, I opened the spiral notebook to the first page. My whole body vibrated with excitement. The miracle of this moment hovered in the room. In bold, black ink, I wrote: "The mathematical structure of the cosmic microwave background." The cosmic microwave background is the name scientists have given the particles of light from our cosmic birth. It had been discovered at Bell Labs in New Jersey just fourteen years earlier, and already mathematicians and physicists from around the planet were working to deepen our understanding of it. The holy grail of mathematical science was the equation synthesizing the general theory of relativity with quantum physics. A first step had been taken with the unification of the electromagnetic interaction and the weak nuclear force. Emboldened by this breakthrough, mathematical physicists were developing theories that would integrate the strong nuclear interaction. The remaining force, gravity, would surely follow.

Breakthroughs in mathematical science often require new forms of mathematics. The classic case is Newton's invention of calculus to explain the motions of the planets around the Sun. The twentieth-century example is Einstein drawing upon the new geometry of Bernhard Riemann to create equations that would explain how light traveled on curved paths and how supermassive stars collapsed into black holes. In my attempt to investigate the origin of time, I too was using new mathematics, KAM singularity theory, the brainchild of the Russian mathematicians Kolmogorov and Arnold, and the German, Moser. No one had thought to use KAM mathematics to this end, and I could be the one to break open the mystery. No doubt it sounds pretentious to admit to such an ambition, but *someone* was going to do it. Why not me? I captured my aim in the notebook with a single phrase, capitalizing each word, "To Understand the Origin of the Universe."

Glowing with satisfaction, I put the pen down and leaned back in my chair. Thrilled by the great journey before me, I gauged the challenge. The creative work would have to take place early in the morning before classes began. Summers would offer time free from interruptions. Both the University of Washington and the University of British Columbia held innovative symposia on mathematical cosmology, as did Stanford University and the University of California at Berkeley. Once again, I outlined the approach I would take, filling page after page with equations, not bothering to prove any of the transitions but simply articulating how I intuited things might go. Later on I could return and see if my guesses could be proven mathematically. For now, it was all intuitive hunches.

AFTER A few hours, my mathematical reflections died down and were replaced by a series of reveries. Dad telling me his grandmother had never gone to school, had never learned to speak English, had never learned to write. Murmurs from teachers came back. Their encouragement. Their confidence. All of these surfaced in my moment of deep joy. I had made it into the world of professional scientists. As I floated in these reveries, I suddenly noticed Denise standing next to me, smiling. She wore a red nightgown that highlighted the reddish tan of her skin. For the last week, she and our two-year-old son, Thomas Ian, had spent every day on the beach or in the rowboat at her parents' cabin on the Puget Sound. The beach cabin was one of the reasons she was so happy the University of

Puget Sound had hired me. For years she had been hoping to return to the Seattle area after our sojourn in California and Oregon. Without saying a word, she plopped down on my lap.

"What a big day for such a little guy!"

As soon as she said the words, the full reality of the moment came rushing back. I was late. Surely the opening session had begun. The day before, the president's secretary had reminded me that new appointees would be introduced first. There was no time to explain, nor to change clothes. I had to get there before the president, while addressing the full assembly, called my name a second and third time without a response. As I rushed out the front door and down the steps to the street, Denise, laughing, called out from the doorway: "You're going to meet some lifelong colleagues. Have *fun*!"

2.

Mathematical Cosmology at the Tacoma Yacht Club

The fall convocation was held at the Tacoma Yacht Club, a few miles to the north. In my white 1964 Simca, I took Cedar Street to Yakima, and with my mind preoccupied with how I would explain my late arrival at the conference, I let the automatic pilot take over, which guided the car to our parish church on North J. After an illegal U-turn, I roared back and dropped down 30th in its long descent to Commencement Bay. Warehouses and machine shops covered the tideflats of the Puyallup River to the south. Industrial clouds glowing a metallic yellow drifted across the bay to Browns Point. Turning north, I raced along the shoreline, passing the seafood restaurant Harbor Lights, with its gray wood planks and twinkling lights still on, then Clinkerdagger, the fancy new kid on the block, which resembled an Elizabethan England public house, and then the defunct Top of the Ocean, a converted ferryboat, which had been my favorite as a child when Dad took us there for all-you-can-eat salmon and strawberry shortcake laid out on a buffet table with a life-size mermaid carved in ice. The once-shining white ferry was now a black, burned-out husk, destroyed by fire the year before. It was rumored that mobsters with tie-ins to Tacoma's law enforcement agencies had torched the place.

The octagonal yacht club sat at the end of a long black slag that extended into Commencement Bay. Half a mile across was forested Vashon Island, my grandmother's birthplace in the nineteenth century. The entire faculty and administration were outdoors on what must have been the coffee break, the men in sports jackets and ties, the women in stylish suits and colorful dresses. As I snaked through the crowd looking for the mathematicians, President Phibbs caught my eye. Dressed in an impeccable gray suit with a flaming-red bowtie, he lifted his hand with a flourish to call me over. I could have changed my clothes, as another ten minutes would have made no difference. But it was now too late to alter anything. My inauguration as a professor of mathematics was fated to take place in blue jeans and Chuck Taylor All-Star basketball shoes.

Phibbs introduced me to Sheldon Schlimmer, the infamous Sheldon Schlimmer. I had never met him, but his name had been mentioned several times when I had interviewed for the position.

Phibbs was jubilant.

"I snagged Sheldon just now on the chance I'd find you. I have been waiting months to introduce you two."

When I first met President Phil Phibbs and heard his vision, I knew the University of Puget Sound would become home for me. His dream was of an ongoing conversation beyond the traditional boundaries that kept the sciences separate from the humanities, an idea I found exciting. A university where all fields of knowledge were in conversation with each other would be a pathway into the subtleties of knowledge. Phibbs was going to transform Puget Sound from its sleepy Methodist origins to a nationally ranked liberal arts college.

I apologized for my late arrival.

"Tosh. You are about to begin a conversation that will change the world." He smiled at his hyperbole. He was naturally theatrical, and with the high energy of opening day, he verged on mania. His eyes and nostrils grew larger with his radiant face. "You two will put the University of Puget Sound on the map." It was this extravagant praise that enabled Phibbs to assemble an impressive array of scholars and scientists. With a polite bow that showed his balding head, he moved off into the crowd, leaving the two of us alone.

WITH AN Olympian apathy for his surroundings, Schlimmer drew a pack of Camels from his coat pocket and stared at the sea with steel-blue eyes.

"I would offer you a cigarette but being from the West Coast, you probably despise the vile habits of European mathematicians."

He was right, I did not smoke, but I took the cigarette anyway with trembling fingers. He proclaimed that the health obsessions in the Pacific Northwest irked him. "Cigarettes might kill us, but nothing else has the taste of pure thought." Born in Berlin, and educated at Brown University and Princeton, he confessed, with surprising frankness, his resentment for having to work in Tacoma, "the wastelands of Western civilization." If not for his wife's intense hatred of the Los Alamos National Laboratory, he would still be working there. He wanted contact only with mathematical colleagues and regarded his teaching responsibilities as "a case of acne that won't go away." When he spoke of his research, he looked me in the eye for the first time. "Theoretical physics is stalled. We have the standard model to eleven significant figures, but the whole edifice reeks. It's a Rube Goldberg contraption that makes Ptolemy's epicycles look like elegant haiku. I'm not alone in this gloomy conclusion, but none of us wants to admit it. We all know how fast the grants would shrivel up. But alone at night, one has to wonder if the whole thing should be scuttled. Something new is needed. My long-range project centers on using Rene Thom's catastrophe theory to unify relativity physics and quantum mechanics."

Inside, I moaned with delight. I nodded *yes yes yes* as he spoke, tried to take a drag from the cigarette, missed my mouth and bent it against my bottom lip. A flood of ideas came rushing in. I did not know what to express first. On my very first day, I had discovered a colleague interested in catastrophe theory, my own specialty within the mathematics of singularities. Schlimmer could become my gateway into the mainline of mathematical cosmology. He had worked at Princeton, home of some of the century's greatest cosmologists. I stuttered as I explained that our research passions overlapped. He looked at me with surprise, as if just discovering I was there. He smiled his big, charming smile.

"You should join our seminar," he said, dropping his cigarette and snuffing it with his shoe. "Thursday mornings at nine."

MY DREAM of penetrating into the truth of the origin of the universe was actually happening.

3.

Primal Light in a Basketball Gymnasium

The day after the autumn faculty convocation, I participated in a panel designed to orient first-year students to the academic world. A handful of professors across disciplines and at different stages of their careers would present a synopsis of their research interests. The panel took place in the University of Puget Sound's Memorial Fieldhouse where the basketball games were played. Four of us professors sat up on the stage behind a long table with a white cloth and pitchers of ice water. A thousand students waited in folding chairs set up on a brown plastic tarp to protect the shiny floor from scratches.

As mentioned above, my particular research focused on the mathematical structure of the cosmic microwave background, the light that had been released in the explosion of the universe's beginning. My plan for the panel was to sketch out some of the current approaches to the radiation and indicate my own particular hypothesis. I had prepared a five-minute summary of my work. As I sat at the table with two other professors, I nervously reviewed my remarks.

Dolores Maro, from the Classics Department, started us off. As she stepped to the microphone, her flamboyant clothing drew everyone's attention: a maroon Renaissance dress with a bright scarlet hat and a pheasant feather that swooped down past her left ear. Stooped forward slightly, she waited at the podium until her charisma quieted the last of the restless students. She began by clearing her throat, then filled the room with a voice powerful enough to reach a large crowd even without electronic amplification.

Professor Maro described the challenge of translating the words of ancient Greek poets such as Hesiod into modern English. A translator needed to penetrate into Hesiod's mind in order to understand what he meant when he wrote of "primal light" or other realities. A successful translation awoke the same feelings in a contemporary human that Hesiod awoke in his fellow Greeks three millennia ago. Hesiod depicted primal light as divine, but such designations were no longer viable in the modern world with its certainty that no Greek divinities inhabit the universe. To say to the modern mind that "light" is divine is to say nothing at all. Maro concluded that Hesiod's beautiful poem lived in a realm beyond the capacities of the modern industrial mind with its materialist fixations. She summarized her point by saying, "To ask a contemporary human to understand Hesiod would be like asking an ant to learn calculus."

After this gloomy conclusion, with a handful of students clapping politely, Jeffrey Bland from the Department of Chemistry took over and enlivened the basketball gymnasium with his extravagant claims concerning the benefits of what he called orthomolecular medicine. But I was too mesmerized by Dolores Maro's remarks to listen. Both by what she had said concerning Hesiod's concern with primal light, and by what she had said concerning the shrunken modern mind.

I WENT last. Though I had been ready to describe several mathematical approaches to the singularities associated with the cosmic microwave background, Professor Maro's talk had conjured something new in me. A feeling, an excitement. As I stood at the microphone, staring at the sea of faces, I dropped my original idea and said that I—like the poet Hesiod—was interested in the light from the beginning of time. I explained that scientists had been studying this light for fourteen years since its discovery in 1964 by Arno Penzias and Robert Wilson working at Bell Labs in Holmdel, New Jersey. Now that we knew how to look, we could examine this primordial light of the universe as it landed on Earth. We had learned to decipher its signals so we could piece the information together like archeologists reassembling stone shards into the statues that once graced the palaces of ancient civilizations. But there was a difference. With us, the shards were not bits of marble but waves of light from the beginning of time.

I became more exuberant as I spoke. I imagined myself out there in their midst, hearing all this for the first time. "It is astonishing," I said. "This light has been traveling toward us for billions of years, and now, for the first time in human history, we can see it as it arrives freshly, here on Earth, after this long, long journey. It is right here in this gymnasium. We are bathed in it, moment by moment. If you cup your hands together as if getting water from a stream, approximately seven thousand photons of light from the dawn of time pass through the space between the palms of your hands each instant.

"Do you see what this means? We are in *physical contact* with the very origin of the universe. I'm not making this up. It's empirically verified. We are the first humans in the history of the world to know how the universe started. Those photons of light left the origin and roared through galaxy after galaxy and trillions of stars and on and on before finally reaching Earth. The primal light is right here. That's what we're studying. The first light, the primal light, the original light from the dawn of time."

MY HEART was beating and my visual field pulsated. This was my first public talk. I had taught mathematics classes as a grad student, but I had never stood in front of so many people in my life. With each sentence I had grown more excited, certain the students would be just as thrilled as I was.

But the silence that greeted me when I finished showed they weren't. It was so surprising. No response. Nothing at all. Awkward and self-conscious, I turned toward my seat. When I got halfway back, the applause began. As it rolled forward to the stage, I sat down next to Professor Maro. I tried to pick up the ice water. My right hand was shaking. I used both hands to carry the glass to my lips. Water spilled down my cheek. Dolores Maro turned to me.

"Was that an act of cunning or confusion?" She touched her nose with her index finger.

"Pardon me?"

"Your metanarrative," she said.

"My what?"

"Science *über alles.* Was it my reference to Hesiod's primal light that ignited your attack?"

I was stunned speechless. I had no idea what she was talking about. For the last six years I had been buried in mathematical equations surrounded by professors and graduate students who lived inside the mathematics. We had social gatherings, yes, but even then the same small groups formed and quickly sank into a festival of mathematical ideas. No one had ever conversed with me using the terms Professor Maro had spoken.

There were no hard feelings on her part. She patted the back of my hand and smiled. "We should have a conversation," she said.

4.

Music of the Spheres at the Grotto Cafe

Within a week of my cosmic background radiation talk, I sent a memo to Professor Maro asking her to coffee. She suggested the Grotto, a small cafe that stood across from the railroad tracks running along the Narrows, the body of water separating Tacoma from the Olympic Peninsula. For decades the restaurant had been the only food establishment in the area, but in the previous year an International House of Pancakes had emerged that blocked the view and left the Grotto in the shadows every evening.

The sky was a solid gray cloud cover that made the sea white. The tide was out, the beach a stretch of brown sand. A small boat with an outboard motor was stranded on land, tilted to one side. Bright green seaweed, flattened in bunches, waited for the water to return. We sat at a narrow table with a linoleum top and plastic salt and pepper shakers. Dolores had fastened her long gray hair with a bright red rubber band and let the ponytail fall down to her waist. The older waitress plunked down two mugs thick enough to bounce on concrete. She poured our coffee from an overcooked glass pot and said she would keep returning for as long as we wanted more.

I began by apologizing. But as to what I was apologizing for, I wasn't entirely clear. That was my first question for her. What had I said that was so objectionable? Before she could answer, I added that whatever it was, it was indeed done out of confusion, not cunning.

"You suffer from unconscious intellectual arrogance," she said. "In your quest for the true cosmogony, you see no need to consult works of philosophy or poetry."

"But that's why I'm here," I said. "I chose Puget Sound because interdisciplinary research is encouraged."

"Is that so?" she asked.

"It was the main topic at my interview. With President Phibbs."

"Ohhh! Then I was wrong about you." She blinked her eyes, smiling. She was laughing at me. "Not just a pretty face. A genuine naïf!"

DOLORES MARO lived and breathed the Greek classics of philosophy and literature. Her mind centered on the current of thought these ancients offered and she explored their ideas by roving through poems and treatises without needing to consult the texts as she unleashed a stream of sparkling insights into Plato or Homer or Aristotle. I took endless notes. As she reveled in the power of these ancient ideas, I marveled over how different I felt in this ancient Greek mind. Our meetings at the Grotto became a regular feature of my new life. At the end of each session, I staggered home and relayed the highlights of what I was learning to Denise. I found myself swimming in a new world.

THE PHILOSOPHICAL orientation that amazed me most came from Pythagoras, a mathematician who invented the word "philosophy" twenty-five hundred years ago. His intuitions brought to life a mood, or a consciousness, or maybe just a feeling that was different from what I experienced in my study of contemporary mathematical cosmology. Dolores quoted Aristotle's summation of Pythagoras's central insight as, *he harmonia ton sphairon*, "the music of the spheres." The instant I heard the phrase, my mind flared. Pythagoras and his colleagues, she explained, had discovered a fundamental relationship between mathematics and the order of the universe, an eerie connection between the inner experience of listening to plucked strings and the mathematical ratios of the string lengths out there in the physical world. Even before a musician would play, a mathematician could examine the lute and say something accurate about what the sound would *feel* like to a listener. If the ratios of the lengths of the frets were rational numbers, a feeling of harmony resulted. If the ratios were not, an experience of dissonance emerged. For these ancient minds, mathematics bridged the outer order of the universe with the inner qualities of mind and soul.

Overjoyed with his unexpected and even bizarre discovery, Pythagoras leapt to the conviction that the cosmos created music as harmonious as the Greek lyre. Everything in the universe worked together to give birth to song, a song utterable in the symbolic form of mathematics. As I reflected on her words, I came to understand an experience I had had years before.

It was a field trip in 1964. The freshmen of Bellarmine Prep were off to see the Pacific Science Center in Seattle. Fr. Phil Clark, our algebra teacher, drove the big blue bus filled with fourteen-year-old boys whose disciplined behavior stemmed from the presence of Fr. Justin Seipp who glared at us as we traveled north on I-5 to Seattle. We parked at the base of the Space Needle. Freed at last, we stumbled off the bus, horsing around and cracking jokes.

Though I wanted to join in, I held back. I had been taken. It happened the instant we reached the white steps of the grand entrance to the center.

The architects had constructed half a dozen massive white buildings that enclosed a spacious network of streams. Knowing that all the knowledge of the universe was just beyond those white walls ignited a thrill within me. We marched single file on a suspended walkway. Beneath us water gushed skyward from a monumental libation vessel. One could imagine the huge hand of a Greek god lifting this for a drink. The path led to the group of towers in the center of the plaza, and I wandered inside one of these. When I stared up, I saw the delicate lattice work at the top of the tower become part of the blue sky. It had touched something in a wordless way when I was child, but now, under the tutelage of Dolores Maro, I think I understood. The architects had created a temple to experience the order in the universe. The blue sky was not a vast amorphous thing, as I had thought. It was structured. The Pacific Science Center was their way for contemplating this sacred geometric structure of the universe. Perhaps that is what Pythagoras was referring to.

BRIMMING WITH energy from the insights streaming from the brilliant mind of Dolores Maro, I began to feel a kinship with the pre-Socratics, especially Pythagoras, the founder of mathematical science. I liked thinking that my work was rooted in an ancient lineage. But were his insights of any value for contemporary scientists? Was there any connection between his early cosmological speculations and our current work of mapping the universe?

5.

A Galloping Beast in the Thompson Hall of Science

I had just finished my lecture on Einstein's special theory of relativity. The mathematical equations for one of his basic ideas, the so-called invariance of the space-time interval, filled the blackboards. I still had twenty minutes to spare. Perhaps I had raced too fast through the details. I tended to over-prepare for this course since it was loaded with some of the best students on campus, including Oona Fitzgerald who had scored a perfect 1600 on her SATs.

We were on the fourth floor of Thompson Hall, which had earned the nickname "the Boeing complex" because of the close relationship the corporation had established with the Departments of Chemistry, Physics, and Mathematics. Over the years, a significant number of professors and students had worked there. The Seattle-based company had funded part of Thompson Hall's construction when the demand to maintain the university's English Gothic architecture had led to extraordinary cost overruns.

I COULD have ended the class right there. My quota of chalk had already been transformed into the mathematical equations I had written out. I dropped the three leftover stubs into the wire-mesh holder at the corner of the blackboard and opened the class for questions. Oona Fitzgerald raised her hand, her round, freckled face beaming. "What's the meaning of life?" she asked. This evoked some tentative laughter, and she smiled as if she might be joking. But after glancing around, she faced me again and waited. It would have been simple enough to avoid her question with a light remark, but I wanted to honor her sincerity. The bit of courage I needed came when I remembered Dr. Barker's response to the same question I myself had asked a few years earlier in my quantum mechanics course. His irritated reply—"Science doesn't deal with meaning"—left me feeling foolish. As if no real scientist would ask such a question. Only an amateurish pretender. Years later, and his words were still with me.

As I leaned back on my desk and reflected on Oona's question, the strangest feeling arose. The students could see I had taken the question to heart. The mood in the room shifted. A tingling grew inside me. It was as if, unknown to me, I had been waiting for this, and yet I felt like a criminal faced with a forbidden act, something that should be avoided but that was too alluring to ignore.

I told the students what I thought was an important truth, that almost none of us knew our true identity. Just as amazing, we forgot that we did not know our true identity. This strange situation came from the tiny worlds in which we lived. We thought of ourselves as Americans or Chinese, as Republicans or Democrats, as believers or atheists. Each of those identities might be true, but each is a secondary truth. There is a deeper truth. We are universe. The universe made us. In a most primordial way, we are cosmological beings.

Then I said it.

"To take this in, you need to ride inside the mathematical symbols."

I did not know what I meant by saying *you need to ride inside the mathematical symbols.* I just said it.

"Begin with the primal light discovered in 1964 by Penzias and Wilson. This light, this cosmic microwave background radiation, arrives here from all directions. We know that each of these photons comes from a place near the origin of the cosmos, so if we trace these particles of light backward we are led to the birthplace of the universe. Which means, since this light comes from all directions, we have discovered our origin in a colossal sphere of light. This colossal sphere, fourteen billion light-years away from us in every direction, is the origin of our universe. And thus the origin of each of us."

I held out my arms as if clutching a gigantic ball.

"We can speculate about what came before this colossal sphere, but I want to stick with the facts physicists have discovered. The empirical evidence points to a time fourteen billion years ago when our universe consisted of a colossal sphere made of light as well as the primal atoms of hydrogen and helium. That colossal sphere transformed itself into the stars and galaxies and everything else in the known universe."

I came to my answer to Oona's question.

"As this sphere moves forward in time, it evolves under the action of expansion and contraction. That is, as the sphere continues to expand, particular subsets are pulled together via the attraction of gravity. This dual action of expansion and contraction set in motion the creativity that has given rise to every existing entity in the universe.

"If you want to know the meaning of life, look at your hand. Energy flows through your skin and bones without which you would freeze to stone. That flow of energy in your hand came from the beginning of time. Your hand grew out of the colossal sphere like a flower rising up from topsoil. No one in the history of humanity knew that the expansion and contraction of the universe transformed primal atoms into stars and galaxies. Nor did any person know the quantum field theory and the general theory of relativity that govern this sphere of light. None of the sages or kings had the slightest notion of any of this, but now we know the mathematical dynamics by which the universe brought itself forth. Those same dynamics are coursing through us. The universe's creativity is happening now. The exact same dynamics are at work. Our bodies churn with creativity rooted in the beginning of time."

I STOPPED. I had worked myself into a trance. The words I had conjured up to explain things boomeranged back on me. In that moment, I felt the simple truth more deeply than I ever had in the past. *I* was the colossal sphere. All of us were. We were rooted in the cosmic microwave radiation. We were the primal atoms speaking of our fourteen-billion-year existence.

Oona Fitzgerald sat in the first row. I did not want to make her feel self-conscious so I avoided looking at her, but now it occurred to me she was Catholic. Did any of this disturb her religious faith? The students watched in silence. I knew something had happened. A strange intuition arose. This universe—held together by mathematical structures—was *breathing* me. But this thought too slipped away.

The ending buzzer pulled me back. My ordinary consciousness reappeared and took control. Students gathered their books, dumped them into their backpacks, and left the classroom. Hadn't the world changed? I felt foolish, embarrassed. Attending to my lecture notes, shuffling them this way and that, I pretended to be too busy to look up.

As the students filed out, Oona came over, smiling. She wore a simple yellow dress.

"I've decided to change my major. Because of your course," she said.

"Really? You mean to physics?"

She nodded.

I was astounded. She was abandoning her musical career? I knew how important music was for her because she had badgered me for a month until I agreed to attend the fall show where she played a violin solo. After the concert, I met her family members, all of them proud of her musical competence and hard work. Had she spoken with her family about this? Was this a good decision? What had I done?

She attempted to say more but stuttered. She stepped toward the door and, turning back, said: "I love this! I would love to learn this. It's so, I don't know . . ."

Students shuffled up and down the hallway behind her. She shook her head and walked to the door. I thought of calling her back but had nothing to say.

THIS EVENT would have not have taken place but for my reflections via Dolores on the "music of the spheres." In one way or another, Dolores's presentation of Pythagoras's idea had created a resonance that connected his ancient insight with contemporary science's concept of a causal cone suffused with quantum fields. The mathematical equations describing the causal cone were the same now as they were before the class began. But the feeling was different. Could this feeling be the "music of the spheres"? Maybe. But it did not feel ethereal like "music." What had surfaced in my experience felt physical. As if I had climbed onto the shoulders of a giant animal. No, not *on* an animal. I *was* the animal. With my eyes pulled wide open, I had become in imagination some kind of beast-universe galloping into the future.

6.

Point Defiance in the Central Basin

The next morning, Saturday, I awoke soaring with energy. I needed to talk with Denise. It was only in conversation with her that I felt any confidence with a new idea. She suggested we go to Owen Beach so I woke up Thomas Ian and fitted him into jeans and his way-too-big yellow parka. In fifteen minutes the three of us were driving to Point Defiance just north of the yacht club where the faculty meeting had been held.

Point Defiance Park had been established in the nineteenth century and had been protected from urban development throughout the twentieth. Tacoma even forced the almighty railroad to respect the park by tunneling underground so that the original temperate rain forest with its sheer cliffs plunging down to the sea would not be disturbed. We ate breakfast at the Boathouse Grill with long tables strong enough to support an elephant, which a photo on the wall proved. The heavily varnished wood was certainly capable of holding our cups of coffee for another couple centuries. Through the tall windows of the octagonal room, sunlight was fractured into a thousand smiles all across the harbor. When we finished, we put Thomas in his stroller and walked along the shoreline. The rain had stripped the pollution from the sky during the night, so the air was filled with the acrid scent of seaweed and bleached driftwood.

The shore's hard sand was covered with a scattering of small pebbles, which captivated Thomas. The early-morning sky was beginning to cloud up again. As we strolled along the edge of the water, Denise gave Thomas the pebbles one at a time to throw into the water. With each throwing movement of his arm, whether the pebble made it to the water or dropped right in front of him, she exclaimed with enthusiasm and he smiled up at her, feeling successful in spite of the lack of any objective achievement.

We found a spot to sit on the sand and I gave her a run-down of what had taken place in class the day before, the ideas we discussed, the students' response. A new insight had awakened me in the middle of the night. The amazing fact was that the students and I had been affected by the primal light that had been bathing us for the entire span of our existence. But we never knew it was present. We were completely ignorant of its existence. But now, because of mathematical science, we were able to listen to the story these ancient photons tell.

"That's why mathematics exists," I said. "And philosophy. And ideas in general. Ideas are how humanity evolves. The ideas of mathematical science enable us to discover the universe, which leads to change. It's exactly what happened with Copernicus. Think of how different contemporary humans are compared to what was going on in the Middle Ages. All of that came from discovering that Earth was a planet spinning around the Sun.

"It's the same now. This new idea of the universe creating itself through time is transforming humanity. It blows my mind. For a hundred thousand years we humans have had the same bodies and brains. You know what that means? The only thing that separates us from our most primitive ancestors are *ideas*! Philosophical and mathematical and technological *ideas.* Isn't that incredible? Humans from the Stone Age are *identical* to us *except for a bunch of ideas.*"

Far offshore a tugboat hauled a barge stacked with containers. By the way it was riding high in the water, the containers must have been empty. The drone of the engine wavered in the light breeze.

"I'm identical to my ancient ancestors in Africa. I have the same central nervous system, the same brain size, the same anatomy, the same biological capacities. And yet they lived in such a different world. There were no roads, no books, no cars, no telephone poles, no grocery stores, no skyscrapers, no television, no libraries, no warehouses, no hospitals. None of them knew they were the end result of fourteen billion years of cosmic development. None of them knew that photons of light from the beginning of the universe were right there, all around them. None of them would have known they were composed of atoms coming from exploding stars.

"The only real difference between them and me is that my consciousness has changed because of mathematical ideas lodged in me. Isn't that incredible? The only difference between us and our most distant ancestors is a *bag of ideas*!"

THOMAS STARTED to cry. He had dropped a pebble before throwing it and was trying to grab it. Denise offered him three new pebbles but, for whatever reason, he was fixated on the dropped pebble. He leaned over, chubby fingers stretching for it, restrained by the safety straps of his stroller. It became a matter of high importance as Denise worked open the plastic clasps, lifted him out, and plopped him on the hard, wet sand to his instant delight. She sat down next to him as he stared hard at the ground. With his hand flattened out like a tiny shovel, he patted the sand. He examined his palm with the tiny crystals of quartz and sand stuck there, studying it as if a millennium would be necessary to fully take it in.

Denise lifted Thomas up onto her lap. She showed him the pebbles in her left hand. Remembering that whole venture, he picked one up and tossed it toward the water.

"My consciousness has changed," she said. "The biggest change was this little guy. When I was pregnant, I would pause in odd moments and think, 'A human being is becoming inside me.' Giving birth to Thomas was one of the deepest experiences of my life. *That* is what changed me. It wasn't an idea. I already knew the idea—mothers give birth. Everyone knows that. I had read it in books and learned it at school and all the rest. But the idea was nothing compared to actually being pregnant. That's when I experienced it in the depths of myself. That's what changed me."

Thomas tossed another pebble, this time reaching the water. As the rings of water separated from his pebble's splash-down in the Salish Sea, he beamed up at his mother. He was in the midst of what seemed to be one of the top one hundred events of his life thus far. Which balanced out the situation. As he radiated his success, I absorbed my mistake. It was obvious. Denise had demolished my hypothesis about a "bag of ideas." I was wrong. It was not ideas alone. It was deep experience that transformed humans.

Something had been torn away. A level of subconscious certainty. I knew the equations of mathematical cosmology, but had I ever actually *experienced* something like music created by the universe? Denise's pregnancy had been transformative. Had I ever been transformed like that? Transformed by an experience of the universe?

7.

Sanctuary for the Chaotic Transition

I began a search for what I identified as Pythagorean thinkers, by which I meant anyone who had a feel for this "music of the spheres." After poking about in ancient and medieval texts, I further refined my search by seeking out, among the general class of Pythagoreans, those thinkers who took mathematical science or at least contemporary science as their fundamental context. Among these, one stood out, the cultural historian William Irwin Thompson who explored a Pythagorean vision of the universe in his many books. To deepen his research, he created the Lindisfarne Association, housed in Manhattan's Church of the Holy Communion, which brought together leading mathematicians, physicists, evolutionary biologists, brain scientists, geneticists, phenomenologists, esoteric philosophers, and world historians. Their aim was to speculate together on whether or not humanity was at the edge of a new planetary culture, one where the meaning of the universe would arise out of an interaction between cutting-edge mathematical science and more ancient mystical insights. I knew the specialized work of one of them, the mathematician Ralph Abraham at the University of California at Santa Cruz, and by extension I assumed the others were also at his world-class level. I burned with envy just imagining what their meetings in New York City must have been like.

ALL OF this seemed far removed from the University of Puget Sound, but that sense of remoteness dropped away when I happened to come upon a notice for a lunch presentation to interested faculty by a member of the Findhorn Community. Findhorn, in northern Scotland, was one of the creative centers Thompson had identified in his writings. Six professors showed up to hear the talk given by Peter Caddy in the basement of the student union building. I was electrified not by his ideas but by learning that a sister institute, named after the Chinook winds, had formed in the Pacific Northwest, on Whidbey Island in the Puget Sound, and that William Irwin Thompson from Lindisfarne would soon be speaking there.

That evening I called up my friend Bruce Bochte and made plans to make the trek to Whidbey to hear Thompson's talk the following month. Bruce and I had met the first week at Santa Clara University at basketball tryouts and quickly became best friends. On road trips to other universities and community colleges, we would fill up the long bus rides with speculations about the meaning and purpose of human life. We had different foundations to our thinking. His was Earth; mine was universe. It eventually came out that these orientations had been established right in the womb in the sense that each of us had been deeply influenced by our mothers. His mom was dedicated to the ecological vision of Henry David Thoreau; mine, to the cosmic vision of Ralph Waldo Emerson. At sporadic moments in our dialogue, we dreamed of inventing an educational center combining Earth and universe, both ecological and cosmological.

ON THE day of Thompsons's talk, standing on the ferry's third deck as we crossed Possession Sound to a Whidbey Island heavily shrouded in mist,

I read out loud from the brochure. How this center had been founded in a lineage of the famous island of Iona in northern Scotland. Just as Iona had become a sanctuary for Western civilization during the European Dark Ages, so too Chinook was designed both to preserve the best spiritual and intellectual impulses of the modern period and to give birth to a spiritual vision strong enough to provide support as the world went through a time of terrible transitions.

Neither Bruce nor I knew what to expect. The journey took us out of our normal sense of things. The ferry worker, wearing a plastic orange vest and sporting a wiry gray beard reaching to his beltline, lifted two fingers as if wishing us farewell as we clanked off the boat and into the fog covering Whidbey Island. I was navigator, using notes I'd scrawled onto a used envelope when I had called for directions. The woman's voice at the other end of the line had apologized for Whidbey's primitive state. Some of the turns onto the backroads were not marked at all. We got lost once and I began to think we would never find our way out of these forests of evergreen trees. I couldn't even point in the direction of the town of Clinton, which would have a pay phone we could use. We decided to turn up every road in search of our destination and in this way came upon the small wooden sign carved to say, WELCOME TO THE CHINOOK LEARNING COMMUNITY.

8.

Three Worlds on Whidbey Island

Chinook consisted of three or four cabins and a larger farmhouse strewn over the rolling green hills surrounded on all sides by forest. The affable director, Fritz Hull, came out of the office to greet us. Irish, curly hair, sparkling eyes, when he went to shake Bruce's hand, his face broke wide open. Fritz was a lifetime baseball fan and was overcome that Bruce Bochte who played for the Mariners, Seattle's professional baseball team, was here on the grounds of Chinook. Fritz offered to give us a personal tour of the center, and when asked what he was particularly interested in, Bruce said, "Do you have a garden?"

The presentation featuring Thompson was held in the large living room of the farmhouse and included two respondents who were seated in front, the esoteric philosopher David Spangler, and a third whose face amazed me. Seated in a wooden chair, framed by the evergreens through the window, was Father Lank O'Connor, who had taught Greek in my high school. The same wide face, the same black hair, recently groomed, the same owlish eyes. The students at Bellarmine Prep held him in the highest regard. For most of us, he was our first encounter with a major intellect. I had not seen him since sophomore year when he left Bellarmine to study theology in California, but rumors of his life had found their way to me. After finishing at Alma College, he had moved to Seattle University where he transformed the role of campus minister into radical orator. At the height of his career, and with great drama, he abandoned both the Jesuits and Seattle University. In the years following his break, I had often wondered what had become of him, but none of my friends knew.

Now I did. And it was a struggle to take it in.

Three worlds within this country of the Pacific Northwest were colliding right here on Whidbey Island. The Jesuit world of Bellarmine Prep and Lank O'Connor, a world reaching back through Aquinas and Augustine to the religious movement coming out of Palestine; my modern scientific world of the University of Puget Sound, which stretched through Einstein and Newton to the lineage of mathematical science the ancient Greeks had begun; and this third world at Chinook, a world unknown to me, but one with enough weight to involve a mind as powerful as Fr. Lank O'Connor's.

In his talk, William Irwin Thompson sketched out a four-part history of humanity. In the beginning, two or three hundred thousand years ago, *Homo sapiens* in the form of hunter-gatherers mastered the power of the symbol. These new humans discovered that the Moon that lit the sky at night could also exist as a whispered word, and that the magical Sun that warmed the world each morning could be captured in a circle carved in stone. Astonished by this emergence of symbolic consciousness, these early humans invented rituals with singing and dancing to celebrate the numinous presence in all entities.

A different life emerged twenty thousand years ago with the discovery of seeds and agriculture. Hunter-gatherers had lived in the same life patterns for a hundred thousand years before making the dramatic decision to tie themselves to a particular piece of land. In the past they had hunted the forests and fished the rivers; now they domesticated animals and forced the rivers to nourish their seeds. After many millennia living in bands of a few dozen people, they settled into villages with populations in the thousands.

The classical era, beginning five thousand years ago, saw the emergence of the theoretical mind, out of which came forth writing, currencies, philosophies, cities, and empires.

The fourth era of the human story, the modern period beginning four hundred years ago, drew upon the power of science and technology to capture Earth's energies in order to shape Earth's life and to guide Earth's evolution. Though many regarded this as the final victory in our long struggle with nature, it led to a breakdown of the ecosystems and a call for a new era of humanity.

The idea that a new form of human being was emerging mesmerized me. Wrapped up in the local concerns of career and family, I had never given any serious attention to the ways *Homo sapiens* had reinvented themselves throughout their history. I wondered if this larger change had something to do with what was taking place with me. Even though the panel agreed that this new era was unprethinkable—that no one could know the specific details of its form—their speculations ignited me, especially as articulated by Fr. O'Connor. He suggested that the materialist and reductionist forms of knowledge production that had dominated the modern period would soon be completed by a new way of knowing.

WHEN THEY finished, I pushed forward to speak with Fr. O'Connor. He gave himself over to each person, meeting their eyes, attentive to what they had to say. I had never seen him without his clerical collar or his black cassock. But even in his short-sleeved blue shirt, he exuded the same weightiness of presence I had known long ago. When it was my turn, I paused before speaking. Even though I never had him as a teacher, I assumed he would know me. He was such a large presence in my life, I could not imagine I was unknown to him. He waited patiently. I introduced myself and said I was at Bellarmine Prep during his regency. Still no recognition. I told him I had been so looking forward to learning Greek from him, that it was a blow for all of us when he left, that Fr. Justin Seipp, the vice principal, had said he was the most intelligent man he had ever met, that Phil Clark had echoed the same in our logic class. I went on like this for a time.

"Oh, yes," he said at last, perhaps realizing I would continue to badger him until he gave in. "Yes, of course I remember you."

Behind me the line of people waiting to speak grew longer. I had no idea how to ask a question beyond throwing a string of words together—*science, consciousness, the evolution of humanity, conceptual knowledge, experience, illumination.* He helped me winnow my inquiry until he knew what I was after. He named the paleontologist and priest Pierre Teilhard de Chardin and the mathematician and cosmologist Alfred North Whitehead. After a pause, he added the evolutionary philosopher Charles Sanders Peirce. He finished by suggesting that if I wanted to explore questions of consciousness and science from the perspective of India, I needed to study Sri Aurobindo.

Whitehead and Peirce were names I had heard. I knew them as evolutionary philosophers with a mystical bent. *Aurobindo* was hardly more than a strange sound. I knew nothing of him. The only name of real significance for me was Pierre Teilhard de Chardin. My physics teacher in high school, Fr. Harold, had introduced us to him. Teilhard had speculated on the overall nature of the universe, and our teacher, a Jesuit like Teilhard, thought we should know his vision. His lecture broke new ground in that it was the first time any one used electronic visual aids in a Bellarmine Prep classroom. In the dark physics lab, we came to the first slide, a group of dots representing the atoms of the early universe. On the second, the same group of dots sported small blue arrows representing the direction toward greater complexity and consciousness, which we were told was the ultimate aim of the universe. The next slides depicted an increase in complexity with images of stars, Earth, and early life-forms as we moved toward the climax.

It was a climax we would never reach. We heard a *whomp* sound and the projector lost power. Fr. Harold, nicknamed "Old Weird Harold" because of his intense love of physics, fiddled with the machine. He called for the lights, but these were dead too. As we sat there in semidarkness, waiting, I heard a muffled sound from my right. My lab partner, Tim Beckman, bent forward to suppress his chuckling. Using a small strip of aluminum bent into a U shape, he had poked this into the outlet and blown the fuse.

I IGNORED O'Connor's suggestion of Teilhard. The fact that I had met his work in high school disqualified him. As I wrote down the other names, I asked my last question.

"Are any of them alive?"

"No." He thought for a moment. I waited, hoping for one last name. "You might contact Arthur Young," he said. "I believe he is still living."

9.

The West Coast Highway

I discovered that this last-mentioned person, Arthur Young, had invented the Bell helicopter, the first to be constructed in America. In addition to his groundbreaking work as a mechanical engineer, he had written two books exploring the evolution of human consciousness, convinced the discoveries pouring in from the sciences demanded a rethinking of our fundamental notions of the universe.

Young lived right in my neighborhood, in a sense. His Institute for the Study of Consciousness was located in Berkeley, California, only a fourteen-hour drive south of the University of Puget Sound. When I learned this, I phoned the institute for any upcoming lectures or courses Young might be giving. A tentative voice fielded my questions and then said, "I am speaking tomorrow night at 6:00 p.m. You're welcome to attend." Thrilled that this was Arthur Young himself, I blurted out: "I'll be there!"

I quickly did the calculations. It was mathematically possible to make it work. My classes met on Monday, Wednesday, and Friday, so if I left early

Thursday morning, barreled down to Young's talk, and returned right away, I could be back late Friday morning to teach both special relativity and the dreaded computer science course. At 3:30 a.m. the next day, I kissed Denise, told her I'd call her on the road, stuffed two peanut butter sandwiches and four apples into a brown Safeway bag, and fired up the Simca. Roaring down the deserted freeway, I made great time, sailing past the green-and-white signs for Lakewood, Chehalis, Portland, Albany, and Eugene. Then I saw the sign for Goshen, and without thinking, I took the exit.

The steep off-ramp led to a dingy, white cafe. A personal shrine. During graduate school in Eugene, I would trek there twice a year, order coffee at the counter, and reflect on the journey to Mexico I had taken with my high school friends, Joe Turner and Jim Hansen, the summer we graduated. We bought jean jackets for the trip. I wrapped a red neckerchief around my neck to complete the look. We had driven I-5 all the way down the West Coast of North America, stopping only in San Francisco and Los Angeles, and then Guaymas and Mazatlán, before reaching Mexico City itself. When we were drinking coffee and smoking Winstons in the Goshen cafe, Joe declared he would win the Nobel Prize in literature. I said I would win one too, in mathematics, unaware there was no Nobel Prize for mathematics. Jim, pressed to match us, promised to replace Freud as the greatest psychologist in Western civilization.

But now on my trip to hear Arthur Young, I went straight to the phone booth next to the cafe. I pushed a dime into the slot and gave the operator my instructions for a collect call. Together we listened to half a dozen rings. I asked her to call again, and this time we heard Denise's anxious voice.

"Where are you?"

"Goshen," I said. "At my cafe, you know? How are you, honey?"

"Wait," she said. "You're *where*?"

"Goshen, you know the cafe—"

"You're in *Oregon*? Why are you in Oregon?"

"To hear that lecture. We talked about it."

"When?"

"Right before I left. You don't remember?"

Neither of us spoke for a moment.

"That was the middle of the night," she said.

"You seemed awake. You said to leave the details on a note. I left it by the front door."

I heard her put the phone down. A moment later she said, "You're driving a thousand miles for one lecture?"

"Yeah," I said.

"Berkeley? That's where you're headed?"

"I know it's a bit extreme."

"You're always dashing off like this."

"Not *always*."

"The Feynman talk. You roar up to British Columbia and immediately lose your wallet. Or that Russian mathematician in Portland. I *want* you to talk with these scientists. Do you think I don't? Why didn't you tell me before I was asleep?"

"I didn't think I was going. Then I woke up the realized I *had* to go. This is the last time, I promise."

"Will you please be careful. You're not driving back tonight, are you?"

I hesitated.

"Brian, you'll sleep down there, right?"

10.

Sr. Isabelle Mary and Dwarf Stars

During the long drive to Berkeley, I prepared for my meeting with Arthur Young. Many questions beckoned, including one that had gnawed at me since the Whidbey Island talk: "Is there a new form of trans-conceptual knowledge emerging? One rooted in science and yet holistic and experiential?" The feeling I got at the Whidbey meeting was that what they valued most was a kind of body knowledge, things like yoga, emotional therapy, psychoanalysis. It felt as if they wanted to push aside mathematical science. But I wanted both, not just one or the other.

Would Arthur Young agree with me?

SISTER ISABELLE Mary, my second grade teacher at St. Frances Cabrini grade school, had given me a book about stars. It explained that a single tablespoon of a dwarf star weighed fifteen tons. I couldn't get over this fact. Our Chevy station wagon weighed a ton. I imagined squeezing fifteen of these into a tablespoon. That would be a string of station wagons from Lawrence's house at one end of Leona Way all the way down the block. Crushing all those down to something that could fit into my hand would be the weight of one tablespoon of a dwarf star.

A person can understand the number 15 in a purely conceptual way. One can add it to other numbers, multiply it, divide it up into various natural numbers. But the process involving the imagination and a series of crushed station wagons took me beyond the purely conceptual and moved me into some sort of pathway for experiencing a dwarf star. The experience of the dwarf star's density became an early touchstone that altered my basic relationship with the universe. Whether it was dwarf stars, supernovas, or exploding galaxies, the universe contained wonders that drew me toward them. That wonder over the dwarf star was part of the process ushering me into my pursuit of mathematical cosmology. It changed what might be called the mythic structure of my consciousness. I had loved all the myths told to me as a child. I regarded them as true. As a child, I knew that Homer's Cyclops was a real being. I knew that in the ancient world he lived with his kin on a mysterious island in the Mediterranean. But though the Cyclops was real for me, the attraction dropped off. It's not that I thought dwarf stars were real while the Cyclops was fiction. It was simply that dwarf stars had captured my imagination. They fascinated me more than a giant with one eye.

As I raced through the Southern Oregon farmlands, I realized not only that mathematics had entered human experience, but that mathematics led to *new* experiences never felt before. My encounter with the dwarf star was one tiny example of how, in the twentieth century, even seven-year-olds could relate to a star in a manner beyond what their ancestors had known. This insight is not a criticism of early humans. In various ways, primal humans had a wide spectrum of deep experiences of the universe, but they did not have this one. How could they? They knew nothing of a dwarf star's density. Some of them, far back in time, did not have the abstract concept "15." Gigantic advances were needed before humanity could take in the reality of a dwarf star. Mathematics had to be invented. Telescopes had to be constructed and improved. Newton's theory of gravitation had to be envisioned. All of this was necessary before a dwarf star could enter more fully into the mind of a human being.

As I entered the foothills of the Siskiyou Mountains, I flashed on a different view of the scientific endeavor. Science had brought forth an entire mountain range of empirical data. From one perspective, science's mountain had an objective existence in the libraries and data banks of the world. But from another, it was a mountain of potential experience. The vast majority of it had not yet entered into humanity at the level of feelings. But it was beginning. Twentieth-century humans could now directly experience the density of stars. Or the radiance of pulsars. The scientific enterprise had generated the data necessary to feel in a direct way many dimensions of the universe. I reached into my brown paper bag for one of my apples. I had never thought of science in this way. As I munched and continued weaving my way through the forest, this massive Siskiyou mountain became a symbol for the dense experience of a universe hoping to come alive within human consciousness.

Was this the ultimate meaning of science and humanity? To awaken sensibilities and zones of wonderment?

11.

Ralph Waldo Emerson's Cry

Arthur Young ran his institute out of his Berkeley home, a two-story Arts and Crafts design with a long porch shrouded by maple trees and one towering redwood. I arrived at dusk, the house so dark I thought I might have the wrong address. I had read parts of Young's major works on evolution, and on my drive through the endless rain of Northern California, I pondered how to engage with him. Was it presumptuous of me to think one day we might write something together on this new evolutionary cosmology?

The door flew open as if this skinny man with a shock of black hair had been standing there waiting. Before I could introduce myself, he smiled and led me into the entranceway with coats on hooks along the walls and a couple dozen pairs of shoes under them. In the living room, Arthur sat in a wooden chair and lectured his listeners lounging on the leather couches. Cigarette smoke and the sweet smell of pipe tobacco filled the dark room. Thousands of dusty books jammed the shelves and covered all the walls. I had arrived at an intellectual temple where the pursuit of ideas annihilated any concerns for housekeeping.

I approached a woman in the back with black hair cascading to her shoulders and with a snakeskin stretched over her body. As she looked at me, I saw the other half of her head shaved to the scalp, a style that had not yet reached my students in Tacoma. She pointed her finger in the direction of Arthur Young. He waved for me to join him, a small man in his late seventies, hunched over, leathery skin, a few strands of hair combed over his skull.

"I've been looking forward to this," he said as I sat down next to him.

His attention stoked my enthusiasm. He asked me to introduce myself but instead of that I praised his work. I knew next to nothing about his thinking, but I had an idea of his overall aim, which I admired. I thanked him for his courage in taking seriously the idea that the history of the universe was not just "one damn thing after another," but was an unveiling of ultimate meaning. I ended with a proclamation that there was nothing in all of philosophy or theology more important than dealing with this challenge, and Arthur was leading the way. As a follow-up to this fulsome admiration, and perhaps even because of it, Arthur decided to share with us the reason for his success. How as a young man he had made the epochal decision to put off a study of abstract philosophy and focus on engineering and physics in order to learn the true nature of nature.

His first major project was to invent a new flying machine. He sent one design after another aloft, many of them blowing up, and with each experiment the team refined its ideas. He worked in a mad sprint for the glory of being the inventor of the first helicopter. But there was much more at stake. This was the late 1930s, with a looming crisis in Western civilization. Arthur Young's research had strong competition. A group of Nazi engineers was striving toward the same goal of providing its military with the world's most advanced aircraft.

Arthur reveled in retelling the story of his invention of the Bell helicopter. Now and then he wiped the sweat from his upper lip with a handkerchief, his forehead reflecting the dim ceiling lamp. Right in the middle of this story, his wife, who was sitting next to him, stopped the narrative. She made no apologies for interrupting him. He retracted his head like an old turtle resorting to habitual instincts of retreat. As I leaned forward to see her, Arthur tried to scoot back but did not have the necessary quad strength to lift the wooden chair. She addressed me in a voice well practiced in giving commands.

"We are interested in making a grant either to you or your university," she said.

"My wife, Ruth Forbes Young," Arthur said, gesturing toward her.

"What's this?" I asked.

"I want to enlist your services." She leaned toward me, her thinning hair falling forward. "Arthur's theory is what the world needs to escape the hold of materialism. The low level of humanity's consciousness mixed with nuclear weapons is a dire state of affairs. Our question for you is this. Are you in?"

I laughed, confused.

"I'm in," I said. "What does it mean to be in?"

Her face frosted over. She fixed her eyes on her husband to remind him he had come up with this idea.

"Do you know who I am? I am not referring to the Forbes line. My grandmother is Edith Emerson."

THERE ARE psychic equivalents to major physical transitions such as puberty. This was one of them for me. Her words transported me from one world into another. Sitting before me was the granddaughter of Edith Emerson, the favorite daughter of Ralph Waldo Emerson, the most significant philosophical cosmologist America has produced. I could not hide my amazement. Emerson's reflections on the nature of the universe had shaped my life in important ways, but his most profound effect came from his confidence. He helped me overcome my sense of inferiority for having grown up someplace other than Europe. Throughout my schooling, whenever the day's lesson addressed the deep questions of life, our teachers drew from cultural achievements of thinkers and saints from ancient and medieval Europe. This made perfect sense, since the older civilizations had been creating works of the mind for millennia, while the cultural achievements of my home, the West Coast of North America, all took place yesterday. Fr. Fitts, who taught Western civilization for the sophomore classes, made our second-rate status explicit. He claimed that by the early nineteenth century, only three major areas on Earth had escaped significant contact with Western civilization: the immense tropical forests of central Africa, the vast plain of Mongolia, and the mountains and rivers of the Pacific Northwest of North America. He explained this as evidence of the power of Western civilization. We readily accepted the implication that we lived in a philosophical and spiritual backwater. We concluded that if anyone wanted to study the deepest truths of reality, one would find them in the works of Europeans.

Ralph Waldo Emerson countered this conclusion.

His passionate cry lodged in my heart: "Why should not we also enjoy an original relation with the universe?" Something inside me came alive when I read this. Something personal. Couldn't we find our way to an original cosmology? Did we have to creep back to Europe for all our ideas? His words forged a spiritual kinship with me. On some subconscious level, I placed him in my spiritual pantheon alongside Homer, Einstein, Isaiah, and Jesus. My adolescent mind lumped all of them together in the same historical realm, which I regarded as being far back in time and far away in space.

But her words collapsed those temporal and spatial distances. Sitting next to me in the smoke of the Institute for the Study of Consciousness was Emerson's great-granddaughter, now carrying forward his work. This meant for me that humanity's philosophical enterprise was no longer far, far away. It was right here. Emerson's direct descendent was inviting me into his lineage. In this twist of fate, I found myself inside a larger story—that of the American development of cosmology issuing from Ralph Waldo Emerson who urged us to penetrate through the dross and discover the rich and fabulous cause for our existence. The struggle, the hope, the triumph that Emerson had brought forth in his own life, and in the lives of other philosophers and scientists and poets, had led to this moment.

Isn't this what I wanted for my life? Isn't this why I went into mathematics? Wasn't my quest identical, to understand the origin and cause of the universe? Wasn't Emerson speaking directly to me when he uttered his famous cry? Wasn't I destined to answer with a bold "*yes*"?

RALPH WALDO Emerson's great-granddaughter continued.

"When I ask if you are 'in,' I am speaking of a race far more significant than Arthur's competition with the Nazi engineers. Everything is up in the air. We are living in a deranged world where nihilism dominates every major state. The contest today is for the next world philosophy. Arthur's spiritual theory must be defended from the criticisms coming from reductionist scientists, and I am dedicating significant financial backing to make this happen."

The evening discussion concluded, Arthur, Ruth, and I worked out the arrangements, right down to where the checks should be sent. I was delirious with these new possibilities. To be invited into the thrilling work of constructing a world philosophy, to be in contact with the major intellectuals they'd already drawn in, to be ensconced in the cosmological tradition Ralph Waldo Emerson had begun. And to be paid to do all this!

12.

Siskiyou Mountains

Ruth and Arthur offered their guest room for the night, but I needed to leave. I had to teach my classes in Tacoma, which was true, but what I didn't say was how desperate I was to get away. My mind was jumbled. Things were off somehow. I'd been straining to maintain my high enthusiasm. Using the directions to I-5 that Arthur had jotted down, I drove up Ashby Avenue in search of a gas station or a late-night restaurant, but this upscale Berkeley neighborhood had been asleep for hours. I headed east on Highway 24 toward massive black hills that loomed ever larger until the freeway dove beneath them in the Caldecott Tunnel. Beyond the mountain range, the town of Orinda had a gas station near the freeway exit, but cars blocked me from pulling off. In frustration, I accelerated to match the California pace and sailed by Lafayette and Walnut Creek. It wasn't until I reached the Benicia–Martinez Bridge that I calmed down. The dark, mile-wide Sacramento River shone with white ripples on its surface. Farther east a hundred destroyers, mothballed ships from World War II, were waiting to be dismantled into scrap metal to be sold to Japan, the country they had been designed to defeat.

At the Vacaville exit, I pulled into the Chevron station, filled two large Styrofoam cups with black coffee and grabbed a package of six bear claws wrapped in cellophane and a green box of NoDoz pills. At the outskirts of Sacramento, I turned north onto I-5, the freeway that, a thousand miles later, would deliver me to Tacoma. I didn't have to make any more decisions.

ARTHUR AND Ruth wanted to fight nihilism, a noble and good aim. The problem was they had convinced themselves Arthur had solved the universe's mysteries. Which was absurd. In their thinking, all that remained was convincing the world of Arthur's vision. To accomplish this task, they needed a helper to correct any of his mathematical errors. That's what they wanted. A connection to a mathematician in a mainstream university who would endorse Arthur's work. My vanity had blinded me. Their attention had made me feel important, real, an insider.

WITH MY right hand, I ripped open the cellophane and gorged myself, cramming the full bear claw in my mouth with the brown sugar crumbling onto my shirt and pants. None of that mattered compared to canceling the deal with Arthur. I flicked the tab on the coffee lid and sipped what tasted like stale cigarettes. Traffic thinned in the flatlands of California. Past the billboards, the land ran smooth for a hundred miles to the Sierra Nevada in the east. Exhausted from the churning thoughts, I sank into a stupor that left only the drone of the engine and the whine of the tires on the concrete road. Dolores had claimed the corporations owned science, which infuriated me. I had argued that science for me was a holy search for the truth. But look at my behavior. How many years would have gone to waste? How long before I realized I had sacrificed my work for a little wad of money?

AHEAD OF me, a boxy truck led the way with one red taillight. The Moon broke through the clouds, shining on the orchards beyond the barbed wire fences, with the crosshatching of row after row of the plum, apricot, and pear trees running off toward the horizon. It made perfect sense to reach out to Arthur Young for dialogue. He had devoted years of his life reflecting on cosmic evolution, and we had a genuine discussion in front of the assembled audience. He had lamented the lack of philosophical meaning in the science courses of American and European universities, praising my plan to overcome this inadequacy. He even suggested that my decision to speak of a meaningful universe from the view of contemporary science strengthened his conviction that Western civilization was evolving beyond the idea that the universe was a series of neutral, meaningless, thermodynamic energy exchanges.

But then the fundamental problem surfaced. Shorn of technicalities, his cosmology followed the scientific account from the big bang up to the present, but he imposed upon science the ideological idea of the fall of humanity. For Young, the meaning of cosmic evolution was the tired old dualistic view that humanity needed to "turn around" from its sinful materialism and head back toward the primal light from which everything emerged. He had even used an idea from my favorite ancient philosopher. Pythagoras worshiped all numbers but some even more so, like 7. The universe was not random. It was controlled by the number 7, in that the first three stages of the universe were about a descent; the last three an ascent; with the middle, the 4, being the turnaround point, the time in which we lived.

I was dismayed by what I heard, but I sat there and said nothing. This was not an original relationship with the universe. Science had discovered cosmic evolution. Why would we return to nonviable religious beliefs? Why did he need to cage our amazing science in stunted categories?

THE ROAD sign indicated I was entering the Siskiyou National Forest. As my headlights cut into the dark forest, I saw patches of snow between the pine trees. I smiled. I realized I had learned something. Something important at the center of all this.

Arthur's approach had extinguished my enthusiasm because all he had done was slap a religious doctrine onto the scientific narrative. For Arthur, the vast universe was a *backdrop* for theology. But the traditional theologies had been articulated before we knew the dynamics of cosmic evolution. In the lowlands of the Klamath Forest, I saw with crystal clarity the mistake. It was a mistake I would avoid. I would not present the universe as a "physical" side show. Nor would I allow anyone to think that fourteen billion years of cosmic evolution was nothing more than a negligible prelude to humanity's appearance. I would shun all those well-traveled paths. The universe was singing a new story of immensity.

A WAVE of enthusiasm swept through me. I had escaped imprisonment. The development of the universe through time was the new foundation.

13.

The Shakespeare of Mathematical Cosmology

The green sign announced the Siskiyou Summit as the highest point on I-5, which implied that the entire West Coast of North America was mine. Sir Isaac Newton's theory of universal gravitation proclaimed that if the drag from the air could be eliminated along with all the friction from the tires and ball bearings, I could glide all the way to Tacoma with no gas. In fact, I would reach Tacoma with such velocity I would fly through southern British Columbia and up the Fraser River canyon to the highway's end, several miles north of Lillooet. The same would hold true if I turned my car around and headed south through California's Central Valley, continuing down the Baja California Peninsula to the end point at Cabo San Lucas.

To feel the power of this gravitational force, I shifted into neutral. As Earth pulled the car down the mountain, a human being inside reflected on the cosmic dynamics at work. I conjured the forested hills in Southern Oregon and saw my little car soaring up and over them in response to Earth's gravity. This imaginary journey was rooted in the work of Galileo and Newton. Their theories had found their way into modern consciousness so that even with something as ordinary as driving on a freeway, humans could understand their actions as congruent with the processes of the universe.

It will be the same with our discovery of a time-developmental universe, a universe that develops through time from plasma to galaxies to living planets to human consciousness. We will witness our minds restructuring themselves as we learn to think and live in alignment with universe creativity. Feelings of expansion washed over me as I took this in. How eerie that at distinct moments in the twentieth century, the dynamics of cosmogenesis began to surface in the human imagination. Our universe had been creating itself for billions of years and suddenly, through the work of a handful of human beings, the universe found a way to reflect on itself, on how it had developed over billions of years.

Who were those humans who enabled this awareness? Who were the key scientists who became the eyes that saw cosmic evolution? As I sailed through the night, my mind sifted through its knowledge with the aim of naming them.

ALBERT EINSTEIN would be the first candidate for primary discoverer of the development of the universe. His field equations, published in 1916, predicted the cosmic expansion and became the foundation for mathematical cosmologists around the planet. Indeed, his sixteen partial differential equations can be considered the theoretical core of the new evolutionary cosmology. But as significant as that achievement might be, there are problems with choosing Einstein as the fountainhead. Einstein flatly opposed the idea that the universe had an origin in time. Do we want to name Einstein as the discoverer of the expanding universe when he himself insisted, for a time, that the universe as a whole did not change?

If Einstein is not the primary discoverer, the next contender would be the Russian mathematician Alexander Friedman. It was Friedman who tried to convince Einstein that his equations contained the secret of a universe expanding. Even by 1922, he could show that Einstein's field equations allowed three distinctly different worlds, each with a different mathematical curvature. One of these three was the model of a universe expanding throughout time. But Friedman had no way of deciding which of his mathematical worlds matched reality.

To settle the question of the universe's curvature, scientists needed direct evidence. An experiment had to be devised by someone. And that someone was the observational cosmologist Edwin Hubble, who, working in California, gathered the data of a universe of galaxies expanding apart. Hubble was not the first. Vesto Slipher, working in Arizona, the next state over from California, discovered the so-called cosmological redshift more than a decade before Hubble. But Slipher could identify the redshift only because he studied the work of Henrietta Leavitt. Leavitt had found a way to use Cepheid stars to determine the distance from Earth to the stars.

Each of these scientists needs to be included if I was going to honor any one of them. Einstein, Friedman, Leavitt, Hubble, and Slipher. The first two, Einstein and Friedman, provided the theoretical framework of cosmogenesis. The next three, Leavitt and Hubble and Slipher, captured the data. But the real question, the most fundamental question, was this: Who put it all together? That was a question easy to answer.

Georges Lemaître, the Belgian mathematical cosmologist, invented the theory that envisioned the cosmos expanding from a powerful explosion at the beginning of time. His 1931 paper hypothesized that a "primeval atom" had erupted in the distant past and sent matter flying apart. Indeed, it was Lemaître's paper, combined with Hubble's data, that finally convinced Einstein. If only Einstein had seen that his mathematical equations had predicted all this. A bittersweet moment. If he had possessed more confidence in his own abstractions, he might have been the one to make the announcement concerning the grand beginning of everything. Instead of that triumphant declaration, Einstein had to admit defeat, and did so with wonderful courtesy. On the day he and Lemaître visited Edwin Hubble at Mt. Wilson, Einstein summarized the situation with a simple announcement: "Lemaître smashed my idea of a static universe with a hammer blow."

I SAW the gas gauge was at empty. I could probably make it to Medford, but to be sure I pulled off at Ashland, the West Coast's premiere venue for the plays of William Shakespeare. Even though the theater season had come to a close, Elizabethan banners hung from poles down Main Street. Many of the businesses tied themselves to Shakespeare. The Bard's Inn displayed a vacancy sign blinking with red neon lights. A storefront window advertised JULIET'S FINEST, a woman's clothing store. I drove straight through the dark downtown, past the Ashland Hotel, and up the slow grade leading out of the city before I found an open gas station.

When the attendant approached, I brought the window down and smiled upon hearing his British accent. Maybe he was an actor from the Oregon Shakespeare Festival earning extra cash. He got the gas pump going, pulled up the windshield wipers, and squirted from his blue plastic bottle. He used a squeegee long enough to reach across the entire windshield. I wanted to ask him if he had performed in any of the plays, and if so, if he might say a couple lines. Or maybe tell me about his love of Shakespeare that pulled him across an ocean and a continent to the West Coast, just for the opportunity to be in one of the plays. But I said nothing, not wanting to bother him.

FOR MOST of the twentieth century, Shakespeare's plays had been performed throughout late spring, summer, and early fall here in Ashland and in dozens of other cities throughout North America. Not to mention the United Kingdom, Australia, New Zealand, wherever a nation spoke the English language. And in translations in another fifty countries. Stories of Scottish kings, of Italian nobles, of Danish aristocrats. Stories from the history of England. Stories that had worked their way so deeply into the fabric of a planet that four hundred years later, I could find myself in an Oregon gas station, remembering the words of Shakespeare's competitor when Shakespeare died, that his works should never be forgotten for he was "not of an age but for all time."

The day will come when something similar will be said of these six scientists, especially of Georges Lemaître. Though Lemaître did not write in the effulgent iambic pentameter of Shakespeare, his mathematical statements will be remembered for millennia. It took humans two hundred thousand years to see the large-scale dynamics of the universe. Lemaître's awareness of the fundamental mathematical harmony in the expansion of the galaxies enabled humanity to determine where, in an empirical sense, the birthplace of the universe is. The work of Lemaître led scientists to it. After thousands of years wondering over the origin of the universe, we found that trillion-degree event that had blazed with such an intensity we can still sense it, still touch it, now in the form of the cosmic microwave background radiation, the afterglow of the universe's birth.

THROUGH THOSE six humans, the creative universe made its dramatic appearance. Those six are the core scientists who brought this new revelation forth. To honor their work, I would reject all attempts to slap an ideology on top of them. The universe itself would have to tell us what it was about.

AS I continued to wait for the gas tank to fill, I wanted to howl in celebration, but I lacked the freedom to release my joy. Even so, an irrepressible smile made its way through my restraints as the gas attendant handed over my credit card. His most satisfied customer of the week.

I roared off, hardly noticing as I bottomed out on the asphalt.

14.

Barnacles at the Boathouse Grill

Even though I didn't reach Tacoma until seven in the morning, I was too wired to sleep. I made myself a cup of coffee and watched cartoons with Thomas. I was on fire and had to talk. When Denise awoke, the three of us drove down to the Boathouse Grill at Point Defiance. The wind from the ocean had raised thousands of small whitecaps on Commencement Bay. The force of it changed the sky constantly. Clouds began as gigantic castles then turned inside out and became gray mountains separated by dark valleys as they were swept north. As if drawing upon this wind energy, a large colony of seagulls screamed at each other as they hovered above the water, then fought for the fish entrails tossed from the dock by the workers.

Over scrambled eggs and rye toast, I laid out the discovery of an insight from my drive. It came in a flash. As I was tracing out the lineage of the major cosmologists of the twentieth century, I came to see that these six scientists were more than sources of knowledge. *They had become constituents of my mind.*

I struggled to explain myself to Denise.

"Strands of insights from these scientists and from the whole history of mathematics have become structures of my mind. These structures are now *me*. It's so strange to realize. Do you see what I mean? Other people have invented these ideas, which I have absorbed. So these other people have invented *me*. The ideas of Galileo and Newton and Einstein shape my daily perceptions. If you take this further, it means our ancestors constructed all the human minds on the planet today. Even though I regard my mind as my own, the fact is, others have built it. Isn't that amazing?"

With a gloomy face, Denise offered her son another Cheerio. She was looking at the smoke from the pulp mills rising above the tideflats, the source of her ongoing anxiety. We were hoping to have another child, and she was certain the toxins in the water and air would end up inside the embryo. We had lived the previous six years in Eugene, Oregon, a mecca for the ecological movement. Tacoma with its industrial commitments was the opposite.

"Denise?" I said.

"Your insights are fascinating. The whole idea that others have built our minds."

"But what?"

"Can the great scientists of our day protect our water? So life can thrive again?"

She ran through the ecological damage, the slag from the smelting furnaces, the toxins from Dow Chemical, the arsenic from the Asarco company.

"Dr. Karlson thinks we can absorb current levels and even more," I said.

"Does Dow pay him to say that?"

WE GATHERED our belongings, and while I paid the bill, Denise pushed Thomas outside in his stroller. Instead of going to the car, she headed the opposite direction. I caught up and held my father-in-law's big purple-and-gold University of Washington umbrella over all three of us against the rain. She bumped the stroller down the concrete steps to the beach. It was low tide. I was wearing my best shoes and hated the idea of getting them wet. Denise left our son on land at the sea's edge, kicked off her shoes, and stepped into the shallows of the Salish Sea. She reached her hand to the black piling at the water's edge, the nearest of a couple dozen that held up the Boathouse Grill.

"Not one," she said. I didn't know what she meant. I studied the piling, which was painted with black creosote. I searched the sandy bottom of the shallow water. "Barnacles can't live here. Commencement Bay is too poisonous even for barnacles."

I HAD difficulty knowing how to respond to this. I was flying with enthusiasm for my realization that my consciousness had been constructed by ancestors like Descartes, Newton, Einstein, and Dirac. Wasn't that completely amazing? My entire life with its decades of study had led to this breakthrough in understanding. Was there anything more thrilling? I understood it had nothing whatsoever to do with these barnacles, or rather lack of barnacles. And I fully appreciated that barnacles were important, and that drinking toxic water, especially as an embryo, is a horrible idea. But does that mean I am supposed to drop my study of the universe and focus on barnacles? It was obvious we should clean up the oceans and rivers and bays, but there was nothing intellectually interesting about any of that. Is that all I'm good for? Researching interesting items that have nothing to do with real life?

The mud was soft at the edge of the Salish Sea. As I waited for Denise, I could feel myself sinking farther into it, but I knew that trying to hurry her up would be as fruitless as trying to stop the tides. She was a force of nature, which is why I was disturbed by what she was saying. I wanted my work to be valued by her precisely because she *was* a force of nature. Her central concerns were children, babies, holidays, families, life. Did the mathematical structures of the universe's origin stack up to that? No. So what was the point? Was there a way knowledge of the universe's dynamics connects to barnacles, to something real, something that truly mattered in people's lives?

15.

The Initial Singularity of Space-Time in the Boeing Amphitheater

At our weekly seminar on mathematical cosmology, Sheldon suggested we switch to the evening. In response, a few scientists said they would have to drop out, but Sheldon didn't care. He wanted our sessions at night so they could be open ended, so that no one would have to break off to teach or attend meetings. He had developed this habit when he worked at Los Alamos directly after receiving his doctoral degree. Even though he arrived in New Mexico decades after the Manhattan Project, the same regimen established by the original crew persisted. It was why I was happy to go along with working at night. In fact, I was excited by the change. It felt as if, through Sheldon, we were being initiated into a way of doing research that came from one of the most creative ventures in the history of science. The two brilliant leaders, Robert Oppenheimer and Richard Feynman, surely knew something important about the dynamics of creativity.

We chose the Boeing Amphitheater in the Thompson Hall of Science because of its six enormous blackboards—three that would slide up to reveal another three that were part of the wall. We filled them with equations and stood back and examined the whole movement of the mathematics. Sometimes we would argue over the ideas, but a lot of the time we stood in silence, thinking, staring at the equations. If we had an idea, we would try it out on the blackboard, to see where it went. If it turned out to be barren, we would erase it and sit down.

We were exploring the equations that led to a surprising calculation made four years earlier by one of the world's foremost mathematical cosmologists, Stephen Hawking. His paper was a milestone in the history of cosmology. As with the arrival of any radically new idea in science, it ignited a period of wild activity leading to a riot of mathematical speculations. We were intent on joining the fray.

Hawking asked himself a simple question: "We now know that the universe has been expanding for billions of years. Would anything change if the initial expansion rate had been different?" The vast pretension of the question gives a sense of why mathematicians develop an unshakeable certainty that equations deserve reverence. Here is Hawking, one human on a tiny planet spinning about one star out of hundreds of billions in the Milky Way galaxy, which is one galaxy out of a trillion. Yet this little speck has the confidence to speak with authority concerning what the universe would become if he made a tiny change fourteen billion years ago, at the origin. Whence such confidence? This was the Pythagorean breakthrough that asserted numbers were real and mathematics formed the fundamental structures of the universe. This axial assumption led to Western science's conviction that with the magic of mathematics, humans can discover the laws governing the vastness in which we find ourselves.

Hawking focused on the ISST, the "initial singularity of space-time." This "initial singularity" receives its name from a breakdown in Einstein's equations when we use them to travel conceptually back in time. These equations work astonishingly well in predicting various empirical facts concerning the universe, such as the temperature of the ambient background, or the average density of the matter in the universe. But the equations fall apart at a certain point. To say that the equations fall apart is to say the equations soar off toward infinity at one point. They soar off at an exponential rate so that as we push time backward, it approaches a limit to how far back we can go. If we pressed the time parameter to this limit, the equations would predict that matter becomes infinitely dense at one particular point in time.

It is nonsensical to say that matter becomes infinitely dense. We conclude that the equations simply fail at that point. The point at which they fail is the singularity of the equations. It is called the "initial singularity" because there is no singularity that occurs before this one. That is why Hawking and his primary teacher and collaborator, Roger Penrose, identified it as the initial singularity. Furthermore, as we travel conceptually toward this initial singularity, not only do the equations predict that the density of the universe soars beyond all limits, they also predict the space between any two particles moves toward zero. These astounding mathematical results come when we push the equations back to a unique point in time, 13.8 billion years ago. Cosmologists interpret this initial singularity as the beginning of the known universe, the big bang, when all the particles are squeezed together. It is either the beginning of the universe as a whole or the beginning of the special branch of the universe in which we find ourselves. However a person interprets this singularity, it has become a central research focus of some of the most creative scientists working within mathematical cosmology.

HAWKING'S WORK on the ISST explored what would happen if the ISST had been slightly different in various parameters, such as expansion rate. Using the mathematical equations for the fundamental forces in the universe, he mapped out the trajectories of an infinite number of these mathematically distinct universes that made up a theoretical "multiverse." Each of these "parallel universes" was represented by a point in a higher-dimensional mathematical space. The trajectory of any one point represented the evolution of that particular theoretical universe that began with its own expansion rate.

What Hawking discovered, which shocked the physics community, was that our actual universe was mathematically unstable within this higher mathematical world. A physical example of an unstable situation is a marble resting at the top of a smooth hill made of glass. If the marble is perturbed even slightly, it will fall off the crest of the hill and roll away. The slightness of the perturbation has to be emphasized. If we imagine a sub-universe consisting of our balanced marble and a hummingbird a hundred miles away, a single sneeze from its tiny lungs would cause the marble to roll off its perch.

TRANSLATED INTO the dynamics of our universe, Hawking's mathematics showed that the destiny of nearby "parallel universes" would not include life. This was a stunning mathematical result. If a parallel universe has an expansion rate faster than our actual universe, the theoretical parallel universe would not be dense enough to build structures. On the other hand, if the expansion of one of these theoretical universes had a smaller value than our actual universe, its trajectory would end in a single point, a black hole. Particularly remarkable was the delicacy Hawking discovered. The size of a perturbation that would lead to these drastically different outcomes needs to be only one part in ten raised to the sixtieth.

THE OPEN question in cosmology was to explain how the expansion rate of our universe could be so exactly right for the unfolding of life. Three main schools of thought formed: the first focused on a multiverse; the second on design; the third on the inner ordering dynamics of the universe. Before I describe these three approaches, I need to emphasize that all three approaches agree that the expansion of the early universe is extraordinarily elegant. Their differences come in how each explains this mind-stunning elegance.

The first group, the cosmologists of the multiverse school such as Stephen Hawking, explain the exactness of the expansion rate by assuming the theoretical universes in Hawking's mathematics really do exist somewhere, even though we have no empirical evidence of any of these universes.

This method of assuming mathematical equations point to the existence of something never experienced before has a long history. One of the most successful occurrences is Paul Dirac who saw in his equations what looked to be a new form of matter, *antimatter*, which was subsequently detected. If it turns out that the multiverse cosmologists are right, and there are an infinite number of parallel universes, our expansion rate has no explanation. With an infinite number of parallel universes, one of them will have the right rate. That we happen to be in the right one is just a matter of chance.

THE SECOND group, the design scientists, such as William Dembski, pursue the idea that the universe is just right for life because an outside divine agent has designed things that way. These scientists have on their side Isaac Newton, who believed just that. Scientists who favor the design approach regard it as philosophically superior to the multiverse approach because it offers an actual explanation for why the universe is the way it is, whereas multiverse cosmologists can only say, "There is no reason for the elegance. It just happened."

THE THIRD approach, drawing upon the inner ordering dynamics of cosmogenesis, and promoted by scientists such as Ilya Prigogine, postulates that the universe is an agent organizing itself in order to develop greater complexity and consciousness. Thus, the elegant dynamics in the plasma at the beginning of time are regarded as something like a cosmic embryo. In this third perspective, the reason for the expansion rate of the universe is to be found not in the past alone but in the future form. The universe is the way it is at the beginning because its future form demands that it be that way so that it can develop into complexity and consciousness.

HOW WAS I supposed to decide among these three?

16.

Primordial Fire in the Lips

Our seminar was following the pathways opened up by Hawking's trail-blazing work concerning the instability of the ISST. Like Hawking, we had embedded the space-time dynamical systems into a higher-dimensional mathematical space, and we were investigating the stability or instability or semi-stability of the other parameters of the ISST.

So there we sat. It was Sheldon's turn to present, and though I and others had made the odd insights here and there, he was in command, which was common in our power dynamics. With his surprisingly nimble fingers he filled five blackboards with chalky equations before slacking off. His original idea had begun to run out of steam. The scientists scattered about the amphitheater stared at the equations, and a few of us tossed out suggestions for how his idea might be developed. Sheldon stamped out each of them. Soon there was only silence. Sheldon stood at stage left, his right hand on his chin, staring. He said nothing, but we knew he would soon enough. His mind was always teeming with ideas from some secret inner fountain.

After more than fifteen minutes of silent staring, my own mind began to roam. I found myself reflecting on the encompassing event itself. I realized that this moment was what I had dreamt of for so long. It was happening. I was actually working with the mathematics of cosmic evolution. That's what we were doing, right here, each of us absorbed in the equations of Einstein and Hawking, prowling around inside them to see what we could find. This was a moment involving the lineage of mathematics going back through Einstein and Hawking to Newton and Pythagoras. It was this entire mathematical lineage that enabled us to think these mathematical equations.

A shift took place. It came to me that these equations as they existed now in my mind were *the mathematical form of the early universe itself.* This shocked me. I had been thinking of the universe's birth as something that took place a long time ago. Which was certainly true. But here I was, fourteen billion years later, thinking the equations of the expanding primordial plasma. From the perspective of the brain, this meant that the complex flow of electricity in my nervous system, in complicated ways, was somehow related to the equations of the early universe. If, instead of thinking these equations, I had been sleeping or running, the flow of electricity would be different. It was an obvious conclusion. I suddenly realized that the nature of the early universe was shaping the flow of electricity in my nervous system because I happened to be reflecting on the mathematics of the beginning.

I could not contain my excitement.

"Shel Shel Shel!" I extended my arms to both sides. Even though I was loath to break his concentration, it bubbled out. "The instability of the beginning. What it really means is that we are deeply enmeshed in that moment. The slightest alteration of the initial dynamics and the whole thing blows up. Or collapses. That means that in a sense *we* are right there at the beginning. It did not *have* to be that way, but it is that way. Those conditions *then* were profoundly *necessary* for the universe to cause our conditions *now.* Then and now *coalesce.* We are that intimately tied together. And we now know it."

The words roared out of me. I was so excited I said it again: "Theoretically it could have been *different,* but it's *not* different. We live in a universe where the mathematical equations of the beginning are alive in us. If you altered them in any way, we wouldn't even *be* here. We would *never* have come forth. Those conditions at the beginning of time are exactly what they had to be for us to allow the mathematics of the universe's beginning to think inside us."

IF I had spoken in a calm and reasonable way, I might have avoided the unfortunate ending, but I was too hepped up. For those few instants, I had become a node where the mathematical equations of the beginning became aware of themselves. Why wasn't Shel saying anything? Another idea for explaining myself surfaced. Using my index finger and thumb, I surrounded my lips in a small circle. Holding the skin in a small hole, I squeaked out the words.

"I couldn't move my lips if our universe had a different origin. There would be no lips. Whenever my lips move, the dynamics of the fireball are there. My lips are right there in the initial explosion. The fireball is in my lips."

Sheldon looked at me with dead eyes. He was bored. When I finished he turned to his equations. Even then, even when it was completely obvious I should clam up, I couldn't stop myself.

"Shel? Do you agree with anything I've said?"

He spoke without turning from the board.

"You haven't said anything."

The silence that followed made my embarrassment worse. I tried to dispel the awkwardness.

"What? Like this is just philosophy?" I said.

"You're always telling us we need philosophy," Sheldon said. As he continued to stare at the equations, he spoke his final words. "Here's the only philosophy I care to hear about." With his outstretched hand, he touched the blackboard, his thumb slightly smudging the Riemann curvature tensor.

17.

When Newton's Equations Fused with Gravity

Shel's rejection crushed my enthusiasm. After several failed attempts to convince him that what I had experienced was important, I abandoned my efforts altogether. It was not his fault. I myself hardly understood what had happened. Our main difference was the stature of mathematical equations. Shel regarded these as the only things worth talking about. My excitement was nothing more than a subjective blip. That blip, he said, had nothing to do with ultimate reality. But from my perspective, my feelings did in fact pertain to ultimate reality. For instance, I was certain the feelings suffusing me in the amphitheater would change my life. They were not a forgettable side effect. And maybe this was a metaphysical orientation. Maybe the experience of sunshine was just as real as sunshine itself.

THOUGH FROM a distance mathematics can seem dry and abstract, from within mathematics itself, the equations offer a contemplation of what might be called divine qualities. At least that is how the archetypal scientists expressed themselves when they reflected upon their work. Isaac Newton and Johannes Kepler were convinced their mathematics uncovered the order placed into the universe by God. In this ability to discern divine presence, they regarded themselves at the level of the biblical prophets. Both Newton and Kepler arrived at such radical conclusions from the ecstasy of beholding, in a direct way, these hidden harmonies. In such moments, mathematics can unsettle the mind as profoundly as alcohol or psychedelics.

MY FIRST taste of this came in fourth grade at St. Frances Cabrini school. Sr. Rita Maria, standing at full attention less than five feet tall, announced to us fifty pupils that there were more trees on Earth than the number of leaves on any one tree. What this meant, she said in a loud voice, is that, "At least two trees have the exact same number of leaves." She stopped to let this sink in. We stared back, confused. Using big letters, she wrote on the blackboard: "If there are more trees than the number of leaves on any tree, there must be two trees with exactly the same number of leaves." She told us we needed to think hard. If we did so, we would come to understand. We would not have to count the leaves on any tree. She provided strong motivation when she asserted, "God wants you to use your noggins."

When school got out, I biked home bursting with excitement to tell Mom what I had learned. She sat at the round oak table in our kitchen nook enjoying her coffee and cigarette. Her face brightened as I charged through the back door. With my hat and coat soaked through from the rain, I jumped into a pell-mell recount. Even though she nodded as I spoke, I was sure she needed my help in understanding, so I concluded each sentence with, "Do you see?" I was deeply fascinated by the fact that knowledge could be gained by pure thought. In order to convince her how amazing this was, I went theological: "We don't have to go out and count the leaves, do you see? It's the truth. No one can change it. Not even God, do you see? Hunh uhn, Mom. Not even God."

FOR ME, the impact did not relate to trees out there so much as to an illumination taking place within. I had actually experienced something that upended at least some of the foundational convictions of the culture in which I had been raised. What my child's mind took from this experience was the strange proposition that truth, as discoverable by thinking, was stronger than God. Sr. Rita Maria's lesson provided an early glimpse into the power of thought. We pupils knew nothing about the seven-hundred-million-year evolution in the vertebrate line that led to nervous systems capable of logic. Even so, we were *participating* in this invisible power, and it awoke in me a deep reverence for the activity of thinking, and especially for how this mathematical thinking related to the universe.

HUMANS HAVE wrestled for centuries with the relationship between thought and reality. A full understanding continues to elude us. We do not know why something invisible and weightless, such as a mathematical equation, seems to be in control of the stars and galaxies. Scientists begin with the assumption that the underpinnings of the universe are mathematical and then focus on the equations with such intensity that any philosophical wonder is pushed to the margins and soon forgotten. But every now and then wonder surfaces and demands to be attended to, which happened to me when Dr. Barker gave a homework assignment sophomore year at Santa Clara University concerning what was taking place in the region between Mars and Jupiter.

The astronomical dynamism I was attempting to understand mathematically is known as the Kirkwood gap. This refers to an interesting feature about the asteroids, the small rocky bodies found mainly between the orbits of Mars and Jupiter. Scientists noticed something unusual about the times it took for the asteroids to journey around the Sun, their so-called orbital periods. Each has its own orbital period. The outer range would be the asteroid Pallas, which circles the Sun every 4.6 years. The fastest asteroids have speeds close to that of Eros, which requires only 1.8 years to complete its revolution. Altogether, there are more than a million asteroids, each one with its own orbital period.

An enigma now enters the story. One might assume that these orbital periods of a million asteroids would be randomly distributed between 1.8 and 4.6. On the contrary, such is not the case. There are particular regions between 1.8 and 4.6 that are seemingly "off limits," regions where there are far fewer asteroids than elsewhere. These are the so-called gaps. What was going on? Something was *preventing* the asteroids from having particular orbital periods. Something was forcing the asteroids away from those regions. What was it?

The mathematical equations controlling the asteroids have been well known since Newton discovered them. I began my evening with his differential equations written out on a blank page. There was no doubt that these equations governed the asteroids; the difficulty was that I could not see how the equations demanded the gaps. Hour by hour, I employed the logical operations that transformed the equations into different forms, always with the hope that I would find one that showed where these gaps were coming from.

Scientists regard Newton's equations as the bedrock upon which modern science is built because his equations allow us to know the solar system's future positions as it moves forward in time. The technical description is that Newton's equations indicate with precision the way the solar system as a whole forces any particular planet or asteroid to change its velocity second by second. With this data, one can determine where that asteroid will be for the next thousand years. Or the next million. So if there are gaps in the real world of asteroids, those gaps must be mirrored in Newton's fundamental equations.

Following the hint from our professor, Dr. Barker, I eventually came to see how the equations pushed the asteroids out of the forbidden zones. The root cause was Jupiter. Jupiter perturbed the orbits of all the asteroids whose periods were in resonance with its own, which happens when Jupiter's orbital period divided by an asteroid's results in a whole number. The equations showed this to be a destructive resonance, similar to an opera singer hitting a particular note and shattering a wine glass. The glass explodes when the waves from the singer's voice resonate violently with the wine glass's structure. Exactly the same with the dynamics in the solar system. If the complex rhythms of the asteroids' revolutions around the Sun can be likened to the molecular structures of a wine glass, Jupiter fractures the Sun-asteroid patterns by eliminating all those orbits in resonance with its own.

I had been working at my carrel for several hours. The stacked paper curled under my pen's impress. Viewed from without, this was nothing more than an unshaven nineteen-year-old kid stuck under florescent light, his textbook on gravitational dynamics crammed into the corner of the carrel. A mess of black ink scratchings in the spiral notebook, a college sophomore staring at the pages in the notebook, staring at the textbook, staring at the asbestos ceiling. But from the inside, it was altogether different. The world had been forgotten; there was only mathematics. As minutes turned into hours, there came a transition point. As I sat absorbed in the equations, the mathematical equations that I myself had scratched onto the paper *fused* with reality. I don't know how else to say it. The mathematical symbols of gravitation, upon which my mind had been dwelling, *became* the very dynamics operating in the universe.

Then the great moment arrived when they shared their secret. A tiny piece of the complexity of the universe revealed itself to me. I was experiencing the intertwined, mathematical fields that order matter. When the light broke in, I got up and staggered about the library, feeling as massive as an elephant. The speed at which I walked might have been the same as ever, but it felt as if I were not moving. I knew my legs were moving, but I felt rooted in place while the stacks and desks and chairs slid past. I soon ended up in the section housing the cosmology texts. This was my secret crypt. The volumes possessed for me a power similar to that of electrically charged plates. When I touched the potent words on their spines, my fingertips tingled. My favorite was *The Large Scale Structure of Space-Time* by Stephen Hawking and George Ellis. I pulled it off the shelf, opened it at random, and, once again, let my eyes feast on the hieroglyphic Christoffel symbols. On the austere beauty of Riemannian geometry. I reflected on the Einstein–de Sitter metric, the mathematics that structures space-time. Inside these equations were neutron stars and the collapsing galactic clouds. Hidden in the depths were pulsars and rotating black holes. Thousands and millions of our ancestors had created symbols that engendered new symbols that led to the symbols before me. I could feel their presence, as if a silent music radiated off the books and enveloped me.

18.

The Sun in British Columbia

When the mental temper of an era shifts in a big way, the essence of the shift is sometimes captured with a single phrase. The breakthrough into the political freedoms of modern democratic governance was given form as "liberty, equality, and fraternity." First felt by a small minority, this phrase soon became the rally cry that confirmed and amplified the passions of many millions. Another historic example is the irruption of the modern protestant religion. The tangled theological and political battles that confused so many were pushed to the sidelines when Martin Luther uttered his phrase, "salvation by grace alone." It can even happen that the victorious slogan is uttered by someone other than the main architects of the change. Such was the case with the revolution in thought initiated by Charles Darwin and Alfred Russel Wallace. The phrase that won the day, which has been repeated endlessly over the centuries, and which neither of them composed, was simply "survival of the fittest."

Something similar might be happening with the change from the fixed cosmos to the new understanding of an expanding and developing cosmogenesis. One of the most attractive phrases yet articulated concerning this new understanding does not come from Albert Einstein, founder of the theory of relativity, or Paul Dirac, founder of quantum field theory, but rather from Freeman Dyson, a colleague of Einstein's at Princeton's Institute for Advanced Study. Here is how Dyson put it: "In some sense, the universe must have known, from the beginning, that we were coming." As one of the central physicists of the twentieth century, Freeman Dyson was offering a new interpretation of the evolving universe that resonated strongly with what I had experienced in the Boeing Amphitheater.

Through a lucky fluke, I had the great good fortune to question Dyson directly about the deep nature of cosmic evolution in a twenty-minute face-to-face conversation. Our encounter came at a conference on mathematical cosmology at the University of British Columbia in Vancouver. As Dyson was a world-famous scientist and I an unknown only a year and half out of graduate school, approaching him was a challenge. At the various sessions and in between the lectures there were always other significant physicists engaged with him, and I was not about to break into their conversations.

I had all but given up when things took a turn at a social gathering of the sort that often follows the hard work of a science conference. In the ebb and flow of conversational groupings, Dyson and I ended up sitting next to each other on a gray couch in a room full of physicists at various levels of inebriation. The warm yellow light from the glittering chandelier above us filled the room. The bar was through the door to the left. Across the room, a long wooden table held a dozen plates of hors d'oeuvres. People flowed in and out, many of whom greeted Dyson briefly.

I was intensely aware that I should not scoop up too much of Dyson's time. Here was a man who was a colleague of Einstein's; here was a man who as a student had the moxie to quarrel with Nobelist Paul Adrian Maurice Dirac; here was a man with the genius necessary to see the underlying coherence in the different approaches to quantum field theory. Nevertheless, I had a simple question I was dying to ask, and, emboldened by the alcohol, I knew this was my chance. I must not hurry into it. My question was absolutely elementary, which meant that there was the danger of a quick reply. I didn't want a quick reply. I needed to engage the deeper layers of his mind. I wanted him to really think about my question, especially from the context of my own approach, and I knew I would only have one small opening to ask it.

We started with small talk, as a warm-up. In response to my conversational gambit about his childhood, he eventually spoke of the thrill he experienced when, at the age of five, he calculated the number of atoms in the Sun. He was not flaunting his Mozart-like precocious mind, which performed such remarkable feats. He regarded his mind as just another surprising feature of this universe, and as he told the story I too was drawn into marveling over the fact that some members of this strange species, *Homo sapiens*, were able to glance at the Sun and know, via the mysterious power of number, the actual composition of a star a hundred million miles away.

His calculation of the number of atoms in the Sun was nothing original. Most physicists and chemists either know how many atoms are in the Sun or can calculate the number to within a couple orders of magnitude using only pencil and paper. All in less than a minute. Even so, this needs to be understood as an accomplishment that required the two hundred thousand years of inquiry set in motion by human wonder, the last stages of which were Leucippus and Democritus who, in the fifth and fourth centuries B.C.E., imagined that every being in nature was composed of atoms, followed by John Dalton who in the early nineteenth century measured the weights of various atoms, to Johann Loschmidt who made the first rough estimate of the number of atoms in a given weight of matter, and culminating with Robert Millikan who in 1909 determined the mass of a single electron. These investigations enabled scientists to calculate with high accuracy the number of atoms in any object whatsoever. Precisely because this historic inquiry had been so successful, a young child—twenty years after Millikan's work—could use his tiny little hand to push a pencil across a thin piece of paper and calculate the number of atoms in the Sun.

The image of a five-year-old performing this act of mind danced in my imagination as I listened. Dyson's eyes twinkled as he told the story. He was experiencing all over again that same thrill. Was it the size of the Sun that thrilled him? The vast number of atoms it consisted of? The fact that he could calculate this? The stunning truth that his existence included this unexpected reach into the nature of things? It was all of these.

Flaring out from my imagination was an image of Freeman Dyson as an embryo. A tiny, tiny fleshy speck. The idea that came to me wasn't specific to Dyson—it could have been any human embryo. It was the thought that here you have this tiny embryo, too small to see, and it cooks its chemicals day after day and five years and nine months later can actually determine the number of atoms in stars millions of miles away. The mere thought of it took my breath away. A dot too small to be seen that sprouted fingers, held a pencil, scribbled on paper, and analyzed our Sun into its hydrogen and helium atoms.

The child Dyson was a fractal of the entire human enterprise. Since life's first appearance on Earth there have been billions of unique species, with humanity one of the most recent. From the perspective of geological time, *Homo sapiens* is something like an embryo. A tiny speck called *Homo sapiens* has only just appeared in the womb of the universe, just a tiny, tiny speck in the vast ocean of galaxies. And yet by sprouting fingers and scratching on rock walls and clay tablets, this speck has come to discover the mathematical dynamics of the vast universe. Modern humanity is the five-year-old Dyson. We have only just emerged and already we have captured, in symbols, the creativity that has constructed us. That is the fractal nature of things. The embryo "knew" how to create a nervous system that could analyze the Sun. The universe "knew" how to create a species that could understand the universe. It was just the same, only now on the large scale.

MY MOMENT to speak had come. I was ready to ask my question. I nudged his attention toward several famously strange numbers, numbers like the expansion rate of the universe, numbers like the ratio of the electromagnetic interaction's strength to the strength of gravity. If these numbers were changed, the world sketched out by the altered equations would be incapable of bringing forth life. The reasons for all this were still beyond our scientific understanding.

I started off with an innocuous question.

"Freeman," I said. I felt uncomfortable using his first name. I would have said "Doctor Dyson," but one of the most well-known facts in the world of contemporary theoretical physics is that Dyson never earned a PhD. He seems to have been too brilliant to have wasted his time. Others say he was too arrogant to submit to the process. But the fact is, while working with Hans Bethe, he went so quickly into original research best exemplified by his historic synthesis of the wildly different approaches of Feynman and Schwinger to quantum field theory, that one idea led to another, which led to another, all of them brilliant. He just didn't have time. "Freeman," I said, "do you like the phrase 'anthropic principle' as a way to refer to such things as the fine-tuning of the expansion rate?"

"No," he said flatly. "But your phrase 'fine-tuning' is just as bad."

He smiled as if to say this was just his way of getting to the issue at hand.

"I agree with that," I said. "Such phraseology ruins what we've learned."

He waited in silence, staring at me. I had heard anecdotes about his intense power of concentration, but to be in its presence was unnerving. It felt like I had stumbled into a biological experiment involving an overgrown peregrine falcon who was extremely hungry. After a brief pause, he nodded and I was encouraged to say more.

"As soon as the phrase 'fine-tuning' is used," I said, "people slide right back into a Newtonian mindset of an unmoved mover who has 'fine-tuned' a universe of objects to work as it does. Which misses the whole amazing truth we've discovered. Would you agree?"

His eyes twinkled.

"What do you suggest instead?" he asked.

"I don't know! I was hoping you had the answer!"

Dyson shrugged his shoulders.

"It's early," he said. "We've recently discovered these so-called coincidences. Eventually we'll get a deeper mathematical understanding, and that's when we'll get the English right. Phrases like 'fine-tuning' or 'the anthropic principle' keep us in the world of classical physics. Such phrases are inadequate. The universe has evolved exquisitely enough to bring forth life; and life has evolved exquisitely enough to understand how it all happened. *That's* what needs explanation. I summarize the situation by saying that the universe—*in some sense*—must have known we were coming. In a scientific way, not just poetic conceit. The universe *knew*. Somehow. To dig into the situation and to determine exactly what it means to say 'the universe knew' is our challenge. Is this what you're working on?" he asked.

I sidestepped his question. In his presence I was too shy to admit to such an ambition. I tried to make up for this evasion by speaking rapidly of a related issue that might enable me to ask the question I was desperate to ask.

"The problem is we think of ourselves sitting here while talking about objects over there," I said. "In a sense that's true, but it's like when you calculated the number of hydrogen and helium atoms in the Sun, a lot of those ideas from Dalton and of course Democritus came before our understanding of the big bang, so naturally they're going to carry some of the older philosophical suppositions into the calculations—well, not the calculations themselves—but into our language expressing the results of the calculations. And the crucial difference, I think—I'm not claiming this idea is original with me, or maybe it is, I don't know, that's one of the reasons I am asking you—the crucial difference is that, like, take the five-year-old Freeman Dyson calculating the atoms. I mean, now we know something. We know that this five-year-old *emerged* from the solar system. That's the amazing discovery—"

I stopped. One of the physicists at the conference was leaning his great domed head down to speak into Dyson's ear. Dyson's face relaxed as he listened. I had noticed lines in Dyson's forehead deepening as I was speaking with him, and I couldn't tell if this was a reflection of his thinking process or an indication that he was irritated with my questions and felt trapped and frustrated. Dyson's face brightened with a radiant smile when the other physicist finished his story.

Dyson turned back to me.

"Sorry, you were saying?"

"I'm sure I've gone on too long," I said. "It's just that, when we think of the whole event of a five-year-old human who has been assembled out of Earth's crust and is now calculating the number of atoms in the Sun, we tend to leave the Sun out of it."

"What are you saying?" He blinked, trying to bring the conversation back into focus.

I had a sinking feeling. I pushed on.

"When assigning credit to this whole process," I said weakly. "I mean when trying to assess who is responsible for the calculation, we tend to forget the Sun, because after all, the Sun . . ."

His smile stopped me. A big happy clown smile transformed all his facial creases into pure mirth. "The Sun is just a big ball of gas!"

He was enjoying a laugh with this naïf sitting beside him on the couch. My face-to-face encounter was suddenly done. I had so butchered the presentation of my idea that Dyson was left thinking it was nothing more than jest. But even as I churned with misgivings about what had just transpired, I knew it was not entirely my fault. I had not explained myself well. But Dyson seemed to have backed away from his cosmological insight that the universe must have known we were coming. It was all murky with emotion. I was too flustered and embarrassed to know what to think or what to say. I stood up to make room for others who were waiting to speak with him. I muttered my gratitude for his time and edged out of the crowded room.

19.

Freeman Dyson's Intelligent Cosmos Just South of the Border

I was on the road before sunrise to make it through the U.S.-Canada border without having to wait hours in the long lines swollen by the recent paranoia over drug trafficking. Making it especially tense was a lifetime of traveling back and forth across the border with Dad. Born in British Columbia, at Xaxli'p, half a day's ride north of Vancouver, he had been given American citizenship when he joined the U.S. Marines to fight in World War II. Something irregular happened concerning his paperwork, I never learned what. Every time we approached the border to visit our relatives or, in the other direction, to return home to Lakewood, he froze with a quiet panic. Once in my teen years I asked him why he became silent when we approached the border. He told me in staccato half sentences that he was working out which story to tell them.

I stayed on Marine Drive, which soon paralleled the Fraser River. When I spied the river, the dark face of my grandmother, Amy Fraser, came to mind. She had taken the river for her last name. When she told me, I never thought to ask her what her real name was. I was just so impressed that she had done this daring act. When I was four years old, the Fraser was the most powerful force in the world. More powerful than thunder or lightning. I was terrified and fascinated by it. Maybe I could take Fraser as my name too. If Dad were here, we would shoot up 91 to visit her in her cramped apartment in New Westminster. I thought briefly of doing the same, but I told myself it was so early in the morning and she might still be sleeping. So with this rationalization I stayed on 99, zipping past Surrey and White Rock to get to the border. There were almost no lines at all, so I stopped at the duty-free store and bought a large jug of bourbon for Dad, knowing how cheered he would be, thinking it was purchased without paying the usual tax on alcohol.

Once through the border, I drove south along Semiahmoo Bay as it absorbed the incoming tide from the Strait of Juan de Fuca, and followed the freeway east as it headed inland. By Ferndale it was farmlands on both sides as far as the eye could see.

I KNEW I was not as brilliant as Freeman Dyson. Few humans were. But I could see how he had contradicted himself. He had laughed at my idea, but it was laughter that could be turned back on him. When he made his radical declaration that the early universe "must have known, in some sense, we were coming," a conventional modern scientist could have retorted: "But, Dyson, the early universe is *nothing but a bunch of quarks and leptons!*" Such has been the view of mainstream science for over a century. If any other scientist had made the statement about the Sun being nothing but a ball of gas, I would have regarded it as nothing other than a trite rehearsal of how modern scientists view the world. But this was *Dyson*, someone who had imagined his way into a powerful new cosmology; *Dyson*, who in his mathematical investigations had come to the intuition that the universe "*knew*" we were coming; *Dyson,* who had intuited that both knowledge and mind were dimensions of the universe, even the early universe, billions of years before humans emerged.

It was clear evidence of the way in which old theories hold us in their grip even after we have seen more deeply into the universe. Dyson's knee-jerk reaction concerning my question was at odds with his own best intuitions as to how the universe operates. He had laughed out of force of habit, dropping into conventional terms to squeeze some humor out of the moment. Anyway, such a backing away was to be expected when a great scientist breaks through the known structures of the universe and arrives at a more comprehensive understanding. Dyson had, through natural talent and long study, arrived at an original relation with the universe. He had achieved both a way of thinking and a way of seeing that went boldly beyond the contemporary limits of scientific understanding.

It's difficult for anyone to stay with a radically new vision of the universe. Convention pulls us back into the old world again and again. It happened with Max Planck, the founder of quantum physics. When Planck discovered the fundamental mathematical equation of quantum mechanics, he was so stunned by what it revealed he tried to talk himself out of it. It was only when other physicists argued with him that he finally came to see that his equation showed us the counterintuitive truths of the quantum realm, that all energy came in discrete chunks.

Perhaps the most heartbreaking example of all such moments of doubt at the highest level of creativity is the one involving Einstein. Out of his original seeing of the universe, he had fashioned the field equations that captured the fundamental dynamism of the macrocosm. But when he saw that these very equations spoke of an expanding universe, he was overwhelmed. At this time there was no evidence at all that the universe was expanding. So Einstein had a choice to make. Either he could accept that he had discovered the fundamental mathematical structure of the universe, that he had discovered—through his mathematical imagination—the ongoing expansion of the universe. Or he could alter his equations, fudge them a bit, take away their radical announcement of expansion, and preserve for himself the traditional view that the universe as a whole is not changing. His fateful decision was to alter the equations in a way that removed their prediction of expansion. But this attempt to hide from the truth came to an end when he and Georges Lemaître traveled to the West Coast and looked through Hubble's telescope and saw the expansion. It was only then that Einstein could understand his decision for what it was—a colossal loss of nerve.

As I drove south, I realized I should have looked Dyson in the eye and implored him: "If I accept your view and assume the universe as a whole 'knew' we were coming, why can't I apply your insight to appropriate *subsets* of the universe? Why can't your vision of things be applied to *galaxies*? Shouldn't we be able to say, following your lead, 'The Milky Way galaxy, in some sense, must have known we were coming?' Or take this another step. Since the Milky Way is composed of billions of star systems, doesn't it follow that this 'knowing' of inherent potentialities could be carried by at least *some* of these star systems, including most obviously our Sun? Wouldn't the Dyson vision of the universe encourage us to investigate whether or not our Sun participates in a differentiated knowing spread throughout the solar system?"

As I continued south on I-5, a single thought absorbed my attention. It was a soaring realization: *We live in an intelligent universe. A universe that "knew" from the very beginning that life was coming.* Our situation was as Dyson described it. At the present time, we have neither the language nor the mathematics to explain how the universe operates. The closest is "intelligence," but the connotations of that word include "brain" and "nervous system." The early universe had neither. Yet even without a brain or a nervous system, the early universe had the capacity to build elegant spiral galaxies and, over time, living beings with brains. Whatever we call this power, it suffused the early universe like a primordial intelligence. The universe, somehow, knew we were coming.

The shock of it filled me with wonder. This was an ancient insight resurrecting in contemporary cosmology. These cumulus clouds, the blue sky, the Douglas firs, these cars in front of me, all arose from some kind of primordial intelligence that gave birth to, but was different from, human intelligence. Though we had yet to develop the concepts that would help us identify it, we were enveloped by it. We had just begun to notice it in our mathematics. We didn't yet know how to think it, how to name it, how to describe it. But we did realize we lived inside it.

WHAT WAS happening in my car was perhaps a microcosm for what was happening to the human species as a whole. Cosmogenesis became scientific fact in 1964 when the cosmic microwave background, the primordial light, was discovered at Bell Labs. Every mathematical and observational cosmologist knew this. But how often are these facts imprisoned in an interpretation that limits cosmogenesis to knowledge about objects out in the universe? What will happen when we turn our consciousness around and realize that our *awareness* of cosmogenesis is also the work of the universe? How will we change when we face the universe and find the universe facing us?

20.

Mount Olympus in the Marysville Denny's

The traffic north of Seattle was clogging up and I wanted breakfast anyway so I looked for a place to stop. I drove until I found the Marysville exit. There was a Shell gas station to the right of the stop sign and a Denny's restaurant just to the west of that. It was a restaurant I knew well. From late high school on, it was one of the spots Joe Turner and I haunted in the hours after midnight, drinking coffee and dreaming our way forward. We were building up the energy necessary to catapult into the future, he as a novelist, I as a mathematician. These sessions devoted to stoking our imaginations took place in three different Denny's restaurants, one in Lakewood, and another in Fife. But it was this Marysville Denny's that had a singular meaning for me.

We had once talked all night long while sitting in one of its booths. We only noticed this fact when in the middle of our escalating dreams some of the faint, gray light of morning oozed through the blinds. This seemed an impossible turn of events. The night, during which our dreaming could be openly pursued and which had led us through entire kingdoms of fabulous, imaginary, heroic adventures, had now been magically replaced. Unbelieving, I had stood up on the black plastic seat in order to reach the controls of the blinds so I could work them open. The world had become gray-black. At the edge of the road was a pay phone booth, and farther on, above the evergreen trees, were the Olympic mountains. Sheer granite cliffs rose straight up into the sky and were bathed in a pink light from an unseen Sun. The peaks were covered with snow and ice. They hovered massively above the Douglas firs. Even in the moment itself, I knew it was as magnificent as anything I would ever experience.

NOW, ON this drive back from Canada, the teenaged waitress who greeted me said I could choose any open seat. I sat down in the same booth I had sat in with Joe. The blinds were open and the Olympics were partially in view, half-covered with clouds drifting in from the ocean. I ordered scrambled eggs and hash browns. And a brown plastic pot of coffee. The waitress wore the same tan uniform as before, but when I studied her face no hint of recognition came back. My moment with the Olympics happened eight or nine years ago. The chances that one of the waitresses had been there when I had climbed up onto the seat to open the blinds were zero; I was only hoping to bring that moment back.

Sitting in the booth, I felt everything linking up. Dreams of becoming a mathematician had deepened here, dreams of peering into the deepest order of the universe. Which was happening right now. Nine years ago, I had been sitting in this very booth, dreaming of finding a way to this moment, now. The connection was so strongly felt, the idea of retrocausality bobbed into my mind. Could it be that this moment now was also present in that moment back then? Was I now drawing that past event toward this moment?

I basked in these fleeting feelings of wonderment. The lamps on the ceiling, the blinds all scrunched up at the top of the windows, the guys at the counter with their backs to me. All of it. The Olympics rising up above the evergreen trees. I didn't know the name of the particular mountain I was looking at. It could be Mount Olympus, the largest of them all. Everything was lashed together. I was overwhelmed that this mountain took its name from the mountain where the Greeks imagined the gods lived.

Maybe they didn't think of the gods like a group of humans. Maybe their stories about gods on Mount Olympus were the best they could do to express their experience of living in a universe of matter suffused with primordial intelligence. I didn't know. What I knew was that their moment was fastened to this moment; their wonder over their gods was connected to my wonder over mathematical order. Did I have any of this in mind in any explicit sense when I drank coffee all night with Joe and marveled at the majesty of the Olympics in the morning light? No. It was not in my mind. But it was there. The dreams back then were woven into this experience now. The equations governing the origin of the universe had ushered in this moment. That's why we drank coffee all night and poured our hearts out. To take the next step on a journey that led to this breakthrough now.

21.

Narrows of the Salish Sea

The new task was to focus our seminar on Dyson's intuition concerning a form of intelligence that permeated the universe. I knew Sheldon would resist. I would have to persuade him that this was an immense opportunity. We would be doing for Dyson what Alexander Friedman had done for Einstein. Our seminar would provide the mathematical bones and sinews for Dyson's hunch that the "universe knew we were coming." Dyson himself could probably get to the mathematical structures before us, but he had lost interest in that line of research. I could tell Shel that maybe the East Coast was too parched to support such a radically new vision of the universe. It was left to us to push it forward by providing the mathematical structures of a cosmic, primordial intelligence. We would be offering the world an original relation with the universe.

THE IDEAL setup for convincing Sheldon of this new approach presented itself when he asked me if I would help him move into his new home out in

Gig Harbor. Normally I would do anything short of outright lying to get out of such drudgery, but as soon as he asked me I knew this was it. He hated any distractions that pulled him away from his mathematics just as much as I did, so he would appreciate my offer. Maybe he would even get excited about the plan.

The morning of moving culminated in the effort of lifting the washer and dryer into the back of his blue Oldsmobile station wagon. It got worse on the drive. As I sat in the front seat, Sheldon explained in breathtaking detail the physiological processes by which the carbon monoxide in the car was mutilating tiny fibrils of our central nervous systems, probably reducing our intelligence by some fraction of a percent. The problem was the window in the back end of his station wagon would not close all the way. The minor vacuum created by the station wagon's movement sucked the exhaust up from the pipe into the car and then the minor vacuum created by our diaphragms sucked the neurotoxins into our lungs. I hated the thought that I was destroying even a tiny part of my brain. To escape the fumes, I rolled down the window to get fresh air, but Sheldon insisted this only made things worse. For now the air inside the car was drawn out of my cracked window and this created an even more powerful sucking action that drew in even more exhaust from the back. I didn't have the emotional strength to demand he stop the car and let me out. So there I was, participating in my degeneration, but at least I knew why. I was doing this to stay in close relationship with Sheldon so I could move forward in my life as a mathematician.

We crossed the Narrows Bridge and took the first exit and drove a mile north. The immense Douglas firs were so impenetrable on each side of the road you could only see a few feet into the tangle of green branches before it was entirely dark. We came to a place where the wall of trees on the right side had been removed entirely, and we turned onto the new dirt driveway. We bumped along the unfinished road until it opened out even farther onto a patch of land that had been gouged by bulldozers to make room for the new house, a three-story affair with a red composition roof and fir shakes still unpainted. The construction crew had only recently finished their work. Stumps of a dozen trees were piled up to one side, waiting to be burned. The naked dirt was uneven with large mud puddles from the recent rain and medium-size boulders that had not yet been cleared away. A wooden plank had been set up in front of the garage and was covered with butcher paper in an attempt to keep the dirt from being tracked inside.

After lugging the appliances up the flagstones and into the garage, which was crowded with brown cardboard moving boxes with "U-Haul" emblazoned in red, Sheldon and I took off our muddy shoes and went into the kitchen from a door in the garage. This would be the moment to spring my idea on him. He pulled open one of the two vertical doors of the refrigerator and poured two glasses of lemonade. It felt contrived to be served by him. It made obvious our overly abstract relationship. Nothing like this had happened before. Equally uncomfortable was his pleasant manner, his smile, none of which was part of our research mode of life.

As I sipped the cold drink and wondered how I would bring up the Dyson idea, I was stunned by the view. The living room ran the entire width of the house with a wall of windows looking across the swirling waters of the Puget Sound. From this perspective, Tacoma disappeared into a dark green forest stretching across the horizon. The evergreen trees bunched up at the shores of the Puget Sound as if they were the front soldiers of a vast army whose ranks began fifty miles east on the slopes of the Cascade Range. Scattered houses could be seen where the trees had been cleared, but it was even yet a continuous temperate rain forest from Point Defiance in the north down to Day Island directly across from us with a dozen fishing boats and scows tied to docks and then, farther south, the town of Steilacoom just visible on the horizon. All of it swooped into me and I felt elevated by this beauty but simultaneously disturbed that this view was not mine.

Only a year before I had been a graduate teaching fellow making ends meet for our little family with a monthly check of $230.76. It came as a jolt to my consciousness that Sheldon, who was only five years older, had somehow managed to buy a place like this. I was flabbergasted. It was especially difficult to imagine because I was still in my graduate student mindset. My one aim was research. I was not blind to the appeal of such a place. Just the opposite. In early childhood I had developed the fantasy of living on a cliff overlooking the ocean somewhere on the lost coast of Washington State. This dream had been ignited when I first stood at the edge of the continent at Cape Flattery and looked out over the Pacific Ocean. The blue-purple-green ocean that had been crashing into the rocks far below for a million years conjured up in my child's mind the image of a place near the edge, just like Sheldon's perch here on the Narrows bluff, a small place where I would be able to think all through the dark winter months.

It had been nothing more than pure fantasy, as far as I was concerned. It had never once occurred to me that such a situation was something I might actually aspire to. At some level of mind there was a wish that such a cabin at the edge of the ocean would appear in my life, but I knew if it were to happen it would be some kind of miracle, a child's wish for paradise. But now I was standing beside Sheldon who had done exactly that, someone the same as myself in many ways, the same education, the same lower-middle-class background, the same mathematical competencies.

I did not bring any of this up but just stood there, mesmerized by the vastness. As I came back from absorption with sea and forest, I made small talk. I asked him what he was like when he was five years old. I wanted him to remember his deepest passion to know the universe even when he was very small. He shrugged his shoulders and ignored me. I pressed him to tell me. He drew the corners of his lips down to express his boredom with the question and said he liked hunting for frogs when his family went to a forest outside of Berlin. In addition to wanting to speak to him about my Dyson plan, I also felt a gnawing urgency to ask him how he could afford to buy this place, but I couldn't bear to prod him so openly. I knew the moment I tried to, the shame of it would make me stutter and I would come across as pathetic. As I kept up the chatter about what he was like as a five-year-old, I could only think of how awkward it would be to introduce my real topic. Instead of going into the Dyson plan, I blurted out:

"You're moving up in the world, Shel." He stood staring out, imperious. "Why the sudden move?"

He shrugged his shoulders. "Boeing offered me a contract."

"Boeing? For what?"

"The Defense Department is dumping a load of moola to ramp up research."

"Okay. What's the work?"

He looked at me before responding.

"I could tell you," he said, "but then I'd have to kill you."

"Seriously?"

"No. Not seriously, Brian." He lifted his eyebrows in disappointment at my obtuseness. "I'm no great believer in secrecy, but one does sign a nondisclosure agreement."

The distance his words created embarrassed me. I cut to a new direction: "Is it related to *our* research?"

He gave me a strange look. Instead of answering my question directly, he swerved the conversation. Now that he would have access to Boeing's ultrafast, state-of-the-art Cray-1 computer, we could run some approximations to see if we were on the right track with our work on the ISST. I nodded dumbly to this suggestion, still taking in his news. Under normal circumstances I would have interrupted with the technical reasons for why this wasn't any great opportunity. We had already proven the trajectories were everywhere dense near the singularity, so no matter how powerful the computer, we would eventually end up with what were basically zeroes in the denominators, which would terminate calculations on any computer, no matter how fast. My dismay prevented any quick response, but as I stood there listening I began to put two and two together. He knew as well as I did where the approximations would end up. This was nothing but a sop.

"Are you leaving Puget Sound?" I asked.

"I would think that's obvious."

"But will you continue with our seminar?"

Shel's face scrunched up as if he had gotten a whiff of rotten eggs.

"You're on your own there," he said. "I'm done working with scrubs."

“You’re okay doing Boeing’s research? That’s what’s going to happen,” I said.

He shrugged his shoulders.

“Where’s the problem? I give them what they want. They give me what I want.”

“Which is what?”

“Gold doubloons. Ever hear of ’em?”

22.

King Tut in Seattle

Thirty-three centuries after his death, King Tut left Egypt to visit America, and one of last his stops was Seattle's Flag Pavilion, located next to the Space Needle. Denise and I stood for over an hour in the long line zigzagging beneath the Needle's high-rising tripod of parabolic support girders.

In the darkened rooms inside, with the soft spotlights highlighting all of the objects—the amulets, pendants, masks, talismans, earrings, necklaces, all the finely wrought works of the ancient Egyptian goldsmiths—Denise rented the audiocassettes to guide her through the museum and was quickly absorbed in the first panel depicting a photograph of the archaeologists poised in front of the recently dug tunnel leading into the pyramid. Freed from conversation, I returned to my thoughts on the initial singularity of space-time. I was presenting the next evening at our seminar, and I was excited by the idea of approaching the ISST with the mathematical insights of Andrey Kolmogorov, the leading Russian mathematician in the fields of dynamical systems and their singularities.

I floated along with the crowd as if I were sleepwalking. I had no knowledge of the objects on display. Out of a sense of cultural obligation, I read some of the dimly lit panels that designated Tut's lineages and the years of his reign and so forth, going through the business of how he was considered the son of Ra, the Egyptian sun god. Delores Maro would certainly regard all such claims as nothing more than mumbo jumbo, her phrase for religion in general and in particular Christianity's claim that Jesus was the son of God and that he had come down from heaven to save us. I had learned early on in our discussions to avoid any mention of religion. It would only ignite another long attack on the intellectual vacuity of all theology "after Kant's *Critique*." None of these matters occupied my mind. As I glanced at the masks and the necklaces, my consciousness fixed itself on the equations that dealt with mathematical singularities.

For the last several days, I had felt myself move into a state that is sometimes referred to as a kind of pregnancy. I was feeling the first glimmers of an insight that would bring peace to my mind. There was nothing to do except continue thinking. It could not be forced. It was outside conscious control. It involved both thinking and waiting. An active concentration on the equations using paper and pencil, and simultaneously a passive tending to the vagueness out of which these symbols arose. Even while using my hands to do my thinking, I waited for the random bit of grit that would land in the supersaturated solution of my mind and ignite the chemical reactions leading to a beautiful crystal.

In that state, I drifted frictionlessly forward through the entire exhibit and found myself in the last room with the green-lit exit sign on the far wall. I would wait here as Denise strolled through the earlier halls. After rapidly glancing around at the displays, I chose a minor piece off to the side of the main flow of traffic so that I could hide unobtrusively and continue thinking of the ISST. The object inside the display case was a small box used for carrying valuable items. It was gold-plated, with designs carved into the plates, and there was a sawtooth pattern along all edges of the box.

For anyone watching me, it would have seemed I stared at the object in order to penetrate into the life of ancient Egypt. In actuality, I was inside the mathematics of the initial singularity of space-time, hardly aware of the sense data.

After a period of time, something strange happened. As I was bent over, staring through the glass, the zigzag pattern on the box vibrated. I did not think it actually moved. But it got my attention. I continued staring at it, not moving except to blink my eyes occasionally. But now, as I consciously pushed aside the mathematics to focus entirely on the sawtooth pattern, it continued to feel as if it were vibrating. I knew the pattern out there was not jiggling about. I knew it was just there on the box, a little zigzag pattern. I knew something different was taking place, but I didn't know what it was, and I was completely frozen, not wanting to break the spell.

I said to myself, *It's as if it were alive*. The word *alive* was more accurate than *vibrate*. Obviously, I knew the pattern on the box was not alive; it was nothing but scratchings on a gold plate. So what was alive? Then it came to me. It was so obvious. I was experiencing the pattern that had entered my eyeball. I was not experiencing the pattern as it was out there on that little box. *That* pattern was on the other side of the glass, inside the display case. I was not over there. I was on *my* side of the glass. Only because the patterns had crossed through the glass and entered my eyeballs were they now living in my consciousness. This was exactly what Dolores was saying when she explained the philosophy of Immanuel Kant to me. I was on this side of the glass experiencing the phenomenon. I was not over there with the scratchings. The whole thing amazed me. In order to enter my nervous system, the patterns of light from the scratchings on the box had to first enter my eyeballs, and with each spurt of blood pulsing through my eyes' capillaries, the patterns inside me vibrated. That was how they were *alive*. They were inside the mundane and mysterious power we call our minds. The patterns were an element of my consciousness; but my consciousness was an element of my living body; so yes, indeed, in that sense, the patterns themselves were living, in that they were *components of a living being*.

As if in a lucid dream, I became anxious that my experience would vanish. I wanted to continue the feeling of being in the very center of the process whereby patterns of light were being transformed, moment by moment, into elements of experience. Not that I doubted the existence of the pattern on the object out there. Obviously there was something out there on the box. But seven hundred million years of vertebrate evolution were working to transform the scratchings on the box into a pattern in my experience. I was a witness. Staring at what I saw, I said to myself, *I am creating you*. By *you*, I did not mean the pattern on the box. I was not creating that. And by *I*, I did not mean Brian Thomas Swimme. My small self was aware that it was only witnessing the transformation. The universe had poured its creativity into the construction of the gold box, the African continent, and my body. My fresh experience of the zigzag was rising up from the deep powers of fourteen billion years of cosmic evolution.

THE EVENT reached its climax when I noticed that one of the zigzags was slightly "off." In that moment I realized something: *This zigzag pattern was constructed by some Egyptian artisan*. Some guy. Some guy with a family. Or maybe he was a slave. In any event, he was a particular person, and this *thing* that was living inside me, that I was contemplating, had been constructed by him and was now placed inside my awareness. A network of individuals was involved—the archeologists who discovered the tomb of King Tut, the shipping company that flew the art objects to the North American continent, the museum curator who had, with velvet gloves, placed the box on its purple cushion, the photons of light traveling along their strange pathways from inside the display case to me. All of these intermediaries were at work here and now, above all the artisan who had carved these "slightly off" zigzag patterns. He was not located back in ancient Egypt. He was present. I had just missed seeing his hand carve the gold plate with his tool. I could almost start a conversation with him. Ask him his view of things. Like, did he actually believe this ten-year-old Tut was the son of Ra?

The experience was so fresh, so fragile, so interesting, I didn't want to move. I was afraid if I blinked it would all collapse. I had not lost track of the fact that I was in the Flag Pavilion in Seattle, nor that the year was 1978, but at the very same time, everybody was there with me, the Egyptian artisan, Tut, Tut's little golden box. They were all there in a real way, newly emerging. It was certainly my experience but, simultaneously, it was their experience too. That's how it felt. It was not mine alone. They were all there experiencing themselves, but now as me.

Somehow or other, this experience related to the origin of the universe. I felt it. I had entered the museum with all the normal expectations. There would be ancient art objects on display. They would be over there, I would be over here and I would look at them. But that's not what happened. They didn't stay put. They had *become* me.

I FELT off. Disoriented. I was tired of thinking about it. What was I doing? It was uncomfortable. Things were weird, hard to think about.

I made my way through the silent groups of museumgoers, pushed the release bar on the door, and entered a brightly lit museum gift shop filled with shiny objects available for purchase. The counters offered stacks of coffee table books emblazoned with glossy, three-color pictures of King Tut's solemn visage. Like the mist that rises up from ground frost when the December Sun finally floats above the mountains, my disturbed feelings evaporated moment by moment. By the time my eyes adjusted to the intense fluorescent lamps, I was home again.

23.

The Snow of North Tacoma

A month later, the dots connected.

THE RED numbers 12:44 glowed on the clock radio. Past midnight. I was desperately aware of how much I needed sleep, but my mind was a storm. I pulled on black jeans, a gray sweater, and basketball shoes, crept down the hallway, flicked on the light, and sat down in the wooden chair at the head of the dining room table. I needed to write about what was happening so I could figure out what to do. Looking up, my reflection stared back at me from a place outside the window. If there were people out late walking on the sidewalk, they would be able to look in at me. It was uncomfortable thinking someone might be studying me as I sat there, so I got up and turned off the light and sat back down.

As my eyes adjusted to the dark, I saw a light flutter of snowflakes drifting down in the soft glow of the streetlamp. The first snow of the season. I pulled off my Chuck Taylors and replaced them with Red Wing boots from the hall closet. Sliding the row of clothes a few inches to the left, I found my dark blue parka. On the shelf above the rack and underneath some umbrellas and gloves was my winter hat made out of sheepskin and with a broad brim, the back of which could be folded down to protect the neck when the wind kicked up. I grabbed my red muffler and crammed that into the pocket of the parka, just in case. Thus equipped, I left.

The snowfall had begun to dust the streets and sidewalks with a fine covering. No car had yet passed, so when I crossed over Cedar Street each step left the imprint of my boot. The dim light from Taranovski's upstairs room lit up the small flakes of snow. He was probably up there, bent over one of his thick tomes. The sidewalk underneath the maple tree in front of his house was still dry. As I walked north to 21st Street, I could hear the swish of tires rising and falling as solitary cars passed by going from east to west or west to east. I didn't want to walk on 21st with its bright overhanging streetlights, so I crossed it quickly and then turned onto C Street.

The houses were dark and the streetlamps dim, so I needed to concentrate to avoid tripping on the upturned edges of the sidewalk. The roots of the elms and chestnut trees had, over the decades, lifted the concrete so that what had originally been flat slabs became a series of gently rising and falling pathways. As I wove through the dark streets, always moving to the north and east, the homes grew gradually larger, but none more than four stories in height. All of them had been built in the last century out of the old-growth fir and cedar of the luxuriant temperate rain forest that had been here. The strength of that wood would last another century. I could feel the grandeur radiating out of their stately architecture. It was in stark contrast to my own upbringing in Lakewood, a half hour south of the city. Our small house with shake siding was inside what remained from the original forest, still sheltering owls hunting mice by moonlight and rainbow trout swimming against the current as Chambers Creek flowed into the Salish Sea.

I walked the streets without direction, desiring only to absorb the night and the snowfall, but suddenly I found myself in a familiar place. One house jumped out. A classic North Tacoma residence, with granite blocks for the ground floor and dark wood for the upper two floors. All of it protected by a black roof with a high pitch and wide overhang for the winter snows that piled up during the occasional big storms. It was dark now but for an unseen lamp in the foyer whose light escaped from the two small windows at the top of the large wooden door. Melinda Smith's home. Melinda, who had asked me to the Tolo dance when she was a sophomore at Aquinas and I a junior at Bellarmine.

Twelve years back, I had entered that very door. Even though I was wearing an ill-fitting black suit because my recent growing spurt had pushed wrists and ankles beyond the lengths of the sleeves and pants, I had been on a high. Things were coming together for me: the results of the National Mathematics Exam had recently been made public, I had made the varsity basketball team as a junior, and beautiful Melinda Smith had invited me to her formal.

Years ago, Melinda's father, Dan Smith, had opened that door. With a laugh, he told me Melinda wasn't ready yet. He himself had just gotten home from his workday and still had his suit on. Mr. Smith was a legendary presence. A formidable corporate lawyer who had been lured from Chicago to Tacoma by the opportunity to join the inner circle of the Weyerhaeuser behemoth, Mr. Smith found a way, even with so many pressing responsibilities, to oversee a seminar on the great books at Bellarmine to which only the highest-performing seniors would be invited. The instant I heard about it as a freshman, I began hoping that one day I would participate in his discussion of the great ideas of Western civilization.

Mr. Smith was drinking a Heidelberg beer from a brown bottle shaped to look like a small keg. In his other hand he held a small cheese and crackers sandwich. He nodded his round, balding head as he studied me. When he took a bite from his sandwich, bits and pieces of crackers fell down onto his protruding stomach. He didn't notice or didn't care if he did notice. As we stood in the foyer waiting for Melinda to descend from the wide wooden staircase, he said he had learned recently that I was interested in mathematics. I assumed this was a reference to the national exam.

"Yes," I said.

"How old are you?" he asked.

"Sixteen."

"Do you know Charles Fefferman? No? Same age. He just graduated from the University of Maryland in mathematics. He's already publishing original research."

With so few words he had canceled out my sense of self. Twelve years separated me from that encounter, but the feelings of nullity returned as fresh as they were back then. The fragility of self-awareness. I had entered the house with an unexamined sense of success and strength. That had been quickly crushed. Using only vibrations in the air between us, he had eliminated what I had subconsciously assumed was the rock-solid foundation of my identity. A painful event, but here it was again in all its power. It was linked to what I felt, but could not name, in the Tut exhibit. It was still nebulous, but I knew it had something to do with identity. With who I was in the universe. In that moment I accepted what was obvious. Our seminar in mathematical cosmology was finished. In Shel's absence, we could not get to the creative edge. No one had mentioned it at our first meeting without him, but it was over. A pillar was gone.

IN THIS quiet neighborhood with no cars in sight, I moved to the middle of the road. The snow crunched under my feet, the squeaky sounds filling the stillness. Even all these years later, as I stepped on the newly fallen snow and felt it squish down under my weight, I remembered how devastated I was that one morning as a child when I woke up and found the snow had gone. Overnight. The white that had blanketed everything had completely disappeared. Mom explained that snow came in flakes made of both water and air. But primarily air. The vast majority of a snowbank was water and air whipped up to appear huge just the same way that sugar and air is whipped up into cotton candy.

It wasn't just snow that melted down into a much smaller volume when heated up. Twentieth-century physicists discovered the same is true of every form of matter. Lead *feels* solid, but when Ernest Rutherford carefully measured the situation, he found that an atom of lead is primarily space, not matter. It's a question of adding up the volumes of the elementary particles making up the block of lead and comparing this to the volume of the block itself. Electrons are point particles with a volume of zero, so ignore those as negligible. The proton radius is a fermi, same as a neutron. In each atom of lead there are two hundred of these protons and neutrons, so the volume of those material particles is like ten to the negative thirteenth of the atom's volume. Heat up any block of lead and it melts down to ten to the negative thirteenth of the original volume.

Tacoma is maybe ten miles on a side, so take all of Tacoma's one hundred square miles and go down a mile into Earth's crust. One mile straight down. That Tacoma slab is one hundred cubic miles. Heat up Tacoma—all these trees, all these lakes and rivers, all these homes, all these roads and stores, all these train tracks and pilings at the sea's edge, all these cars and swing sets and skyscrapers and people, all the rocks and dirt a mile down beneath the city—heat it all up and the structures melt down to the elementary particles. If one put these particles next to each other, the ball they formed would be ten to the negative thirteenth of the original volume. About the size of a basketball. That's all that would be left. The Tacoma slab at a billion degrees melts down to a basketball.

THIS KNOWLEDGE required centuries of brilliant work, but had it changed anything in my life? I knew the ideas, but did I ever live their truth? The universe had taken a basketball of quanta and transformed it into the Tacoma slab. I knew the calculations. But could I experience Tacoma as whipped into existence by the universe?

24.

Dirac's Quantum Field Theory in Knapp's Bar

The one establishment still lit up in this neighborhood nexus of businesses was Knapp's Bar. The entrance was a tall door made of glass so old it looked like running water. The colored lights from within flashed through it. When I pulled the solid door open, warm smoky air with the smell of stale beer poured out into the quiet snowy world. The bartender, in a cool gray collarless shirt and lightly tinted pink glasses, broke off his conversation with the couple at the end of the bar and smiled. I asked for a beer, pulled out a stool, and perched on its brown imitation leather. I put my hat upside down on the empty stool next to me. A magenta light ran diagonally between the rows of colorful liqueurs. The bartender said something to me, but I couldn't hear him over the juke box and I turned my palms up. The small room had a half dozen tiny tables each housing three or four people. Black booths on the far wall were also filled. The wall above the booths was a giant mirror with the same intense magenta bars of light cutting across diagonally. The bartender held an empty glass in one hand, his other hand on one of the four spigots, and lifted his brows in question. I didn't care what brand he gave me. I nodded and he filled the mug and placed it in front of me.

On the napkin beneath my beer, writing as carefully as I could—and even then my pen tore the napkin—I jotted down Paul Dirac's 1928 equation describing the quantum fields of spin 1/2 elementary particles: $i\gamma \cdot \partial\psi = m\psi$. Paul Dirac, mathematical genius of a very high order, one of the fountainheads of the quantum revolution in physics in the twentieth century, had created this equation in Southern England in the early twentieth century, and since that time it has been written down and examined and tested millions of times. It is as fundamental to our understanding of the microcosm as Einstein's field equations are for the macrocosm.

I had imagined asking the ancient Egyptian artisan if he actually thought Tut was the sun god. What if we turned the tables? What if he were the one asking the questions about *my* ultimate belief structures? What authority would I claim?

WOULD I say that God spoke to me about the ultimate nature of things?

Would I tell him that an elder had handed over the village truth to me?

Would I say I had invented a private theory about the world?

No. No. No to all of them.

I would say simply that the universe could now understand itself in a deeper and more subtle way because of Dirac's equation. His equation is a watershed in the history of humanity's investigation of nature. It represents the biggest advance in science since the work of Isaac Newton. Paul Dirac had discovered something profound regarding movement in the universe. Not the movements of planets in the solar system; Newton had done that. Dirac was investigating the movements of elementary particles. He was searching for the mathematical structure that connects their movements the way sinews connect our bones.

Dirac crossed into a new domain of mentality. His knowledge of matter needs to be considered the culmination of many millennia of human reflection. Because of that history of reflection and because of his innate genius, Dirac became the opening into which the dynamics of the universe could *press into* human experience. By his knowledge of matter, and through his profound reflection on the way it interacts, Dirac provided the spaciousness of mind in which the dynamism could reflect upon itself. In that sense, his mathematical equation is a self-portrait of the universe, the universe becoming aware of its foundational dynamism.

In order for Dirac to understand his own work, he had to demolish several structures of belief. He did not understand his own equation because he approached it with false beliefs that hid its significance. But the more he studied it, the more it pointed at strange things beyond the capacity of his modern mind to understand. The initial conundrum concerned energy. Some of the elementary particles in his equation seemed to have *negative* energies, an absurd idea. He feared he had produced nonsense and reworked his derivations, looking for his error. But it was not the equation that was wrong. It was his mind that was wrong.

He rode the mathematical symbols as if they were a magic carpet ride into realms never visited by any of his ancestors. After repeated journeys, and after intense debate with other founders of quantum mechanics, he finally realized the equation was showing him a new form of matter. A form of matter so strange it needed to be designated *antimatter.* These hitherto unsuspected particles possessed a bizarre and troubling feature. If one of them encountered a particle of regular matter, *both would annihilate.* Both would vanish from existence and a photon of light would be created in their stead. Fully aware that this strange idea of antimatter would provoke ridicule, he nevertheless submitted his thinking for publication in 1931. The response was indeed mixed. Many were baffled. Some were put off. One of the founders of quantum mechanics, Werner Heisenberg, was so disturbed by Dirac's insight he admitted to his friends he simply hoped Dirac was wrong; because if Dirac was *not* wrong, things were far stranger than even the founders of quantum mechanics had so far imagined.

One year later, in 1932, Carl Anderson found some antimatter. He allowed cosmic rays to pass through a cloud chamber with a magnet wrapped around the device and studied the paths the particles took. He noticed that one of them curved in exactly the same way as an electron, but in the opposite direction. Having studied Dirac's predictions, he realized he had found the path an anti-electron.

Other particles of antimatter were soon discovered. The anti-proton and the anti-neutron were observed in laboratories on several continents. So antimatter had made its first appearance in human awareness via the singular equation that arose from Dirac's imagination, published in 1928, but not understood by him until 1931. His theory and its verification have altered the history of thought in an irrevocable way. Dirac's imagination destroyed Western civilization's twenty-five-hundred-year belief concerning the eternal nature of elementary bits of matter. The belief that elementary particles formed the foundation of the universe was found to be as false as the flat-Earth theory. Dirac had deconstructed the belief that matter was indestructible. Rather than thinking of matter as a kind of perduring grit, quantum physicists began imagining matter as an excitation, as a flame, as a luminous flash.

His equation showed something more. Rising up in Dirac's reflections was the unnerving suspicion he had discovered the ultimate source of the universe. His equations revealed the existence of an unmanifest realm, which he identified as a kind of "sea." His words morphed over time into the phrase now used in contemporary physics, *quantum field*. Whether depicted by the metaphor of a watery sea or a terrestrial field, this was a realm that created electrons and protons and neutrons. And it was a realm that pervaded the universe. In every cubic inch anywhere in the entire universe, protons and other elementary particles were sprouting forth out of this quantum field, a realm that was untouchable by hand, unseeable by eye. A realm of infinite generativity.

In the year 1931, this sea—this field, this quantum vacuum—which had been operating incessantly throughout the universe for fourteen billion years, suddenly obtained a fitful presence in Paul Dirac's mind. Fitful because this was an advanced and therefore fragile wave that was becoming aware of how space teems with a primordial creativity. Dirac no doubt had profound moments of entering into and actually living in such cosmic creativity. But just as certainly, he would collapse back into the conventional assumptions that space is simply a container in which various objects moved about.

Dirac's theory that particles and antiparticles were incessantly emerging into existence was verified experimentally by Willis Lamb in 1947. Other experiments quickly confirmed the results so that the investigation, which had begun as mathematical speculation, was now established with empirical data. A new understanding had appeared. Dirac's vision was that of an effervescent matter with roots in a primordial potency. Quantum physicists now saw that even the superclusters of galaxies, the largest known structures in the universe, were not inert objects. They were erupting into existence from the generativity of these non-visible, non-touchable quantum fields.

The challenge for entering reality included this massive figure-ground switch, from space as container to space as infinitely creative womb. As these previously unsuspected depths showed themselves to Dirac, as he became convinced that his mathematics had sprung up from the ontological structures of the universe, Dirac, though he had little use for any religion, lapsed into traditional phraseology and announced: "God is a mathematician of very high order." He was not promulgating a belief in God. He was too humble for that. He was pointing to his conviction that an intelligence beyond his own suffused the order in the universe. His equation revealed a dynamic order of the cosmos. In Dirac's awareness, the creativity that shaped the universe was now beholding itself, astonished by what it saw.

THE BARTENDER of Knapp's caught my eye. He glanced down at my empty glass and lifted his eyebrows in question. He had returned to his conversation with the man and woman at the end of the bar. I shook my head. In the loud music, with the purple haze bathing me front and back, I stared at the torn paper napkin with Paul Dirac's mathematical symbols scrawled on it, now a bit damp and torn at the gamma matrix. Dirac's equation, with its Greek and Latin letters, was here in three distinct but interrelated ways—as a flimsy series of scratches on a napkin, as a dynamism taking place in the trillions of particle interactions in this room, and as a pattern my mind employed to make sense of the world.

My head throbbed. The bartender came over and I handed him the first bill from my wallet. He reached into the side pocket of his black apron to give me change, but I gestured with my hand that the rest was for him. He smiled and nodded his head several times in thanks while continuing to gaze into my eyes. Perhaps I had given him a large amount.

He picked up my empty glass, scrunched the napkin, and tossed it into the trash can with a black plastic liner. I stood there in a stupor, floating in the awareness that I too was flashing forth instant by instant. Dirac's equation would now soak up juice from discarded slices of lemon and lime. The interactions of light and matter, the infinite generativity of the quantum fields, these had momentarily achieved their sketchy self-portrait on the napkin. Just a glimpse, and now collapsing back into chaos. I wondered: Would the same happen with my feeling of floating in a sea of potency? Would I toss that out? Would I unknowingly collapse my mind back into thinking of the universe as a collection of objects?

25.

Snowflakes in the Stoplight at Proctor Avenue

Out on the sidewalk I saw that the snowfall had grown heavier. The windshields of the cars parked in front of Knapp's were smothered with fresh powder. I pulled the back flap of my hat down to block the cold air. Using only my left hand, I pulled the red muffler out of my coat pocket and wrapped it twice around my neck. At Proctor and 21st Street the light was red, and although there were no cars in either direction, I stopped out of dutiful habit and stared at it, waiting for it to turn. I sniffed the air. It reminded me of Christmas. When the red light went dark and the green circle lit up below, the light was luminous, surrounded by falling flakes of snow. The lambent green shone directly into my eyes but also arrived reflected by the flakes of snow that swirled gently about the streetlight. Tears welled up in my eyes. Something soft and warm, maybe it was love, swelled within me. Everything was perfect. Denise, Thomas, my family, the snow, the mysterious greenness of the light, my students and colleagues, Tacoma disappearing into a deep sleep under the snowfall, the professorship with the University of Puget Sound, the undeserved gift of a lifetime devoted to stalking the magic of thought and mathematics.

All this joy ignited by nothing more than a traffic lamp with its little metal cup. I was dwelling in fullness. The green light was twinkling inside me after passing through quantum fields that had yawned open and released a flood of quanta. Richard Feynman, one of the greatest scientists of the twentieth century, was so fascinated that the photons knew the optimal pathway. How could they? What did he call it? *Sniffing.* That the light rays knew his calculations even better than he did because they *sniffed out,* from an infinity of possibilities, the unique path that minimized the action. He didn't think photons had noses. It was his gesture toward cosmic intelligence. He used *sniffing* as a finger pointing to a galaxy.

My mind's understanding had been off. This was not a universe of inert objects. That had been my assumption. Before I began to think, I regarded atoms as the rock-hard foundation of the universe. The Tut exhibit had broken that. Tut's box, with its zigzag pattern, had flown through thirty-three centuries to become me. The origin of the universe was fourteen billion years ago, and here, now. Giving birth, even to my lips, even to my experience of my lips, giving birth instant by instant to the universe.

By blinking, I smeared the tears of my eyes into my eyelashes. Now the streetlamp was weeping too. A smile spread over my face. I had taken a small step. I was only in the shallows, but it was all I could take. If there was more, I didn't even want to know about it. I was bursting with all the energy I could handle.

In this universe, one's core identity was a vast community of beings—a traffic light, a box with quirky zigzags, a father of a high school friend, an Egyptian artisan, the expanding plasma. The curators for the King Tut exhibition innocently assumed they had locked their objects in display cases. The laws of the universe would not allow this. It was not possible to fixate any thing. It was not possible to locate one star simply here, another one there. No. The universe was not a conglomeration of inert things. Each entity was a coiled energy about to soar forth to share its being. I could not say what this meant. I only knew that, one way or another, sharing this with the bright young students at the University of Puget Sound was what I was here for.

26.

Galaxies Singing in Different Octaves at Hood Canal

President Phibbs, in his quest to transform the University of Puget Sound into the "Harvard of the West," persuaded the board of trustees to invite professors to propose new and exciting courses for a common core curriculum. The chosen professors would be released from a full semester of teaching in order to develop their new courses, a rare opportunity in a liberal arts university.

To decide which of the proposals would become a "trustee course," the board gathered for three days of deliberation on the shores of Hood Canal, one of the western branches of the Salish Sea. Professors shuttled in and out and made their presentations. My own time slot was 4:00 p.m. on Friday, the last day, and, as I did not want to be late, I sped across the Narrows Bridge and out Highway 16 and arrived at the Alderbrook Resort half an hour early. This included missing the turnoff. Driving beneath the shelter of limbs from the forest on both sides of the road, I overshot the lodge and had to circle back. The sign with its muted lights and carved wood made it clear this place had been designed for the rich.

The curved pathway from the parked cars to the resort had been laid down with small white stones that crunched under my step and were so perfectly round I wondered whether they were natural or synthetic. Two massive logs supported a breezeway to the main entrance. As the double glass doors silently slid apart, and as the older man behind the counter glanced up and smiled, I told myself I should be reviewing my ideas. My mind was too impressed by the personal significance of the event before me. It was my moment of arrival. I may have driven here in my slightly dilapidated car, but no one knew that. I was on the inside. They had *invited* me. Through a lot of work and some good luck, I was now going to address the board of trustees of the University of Puget Sound. Norton Clapp wanted to hear my ideas. It was *Norton Clapp*, chairman of the board, one of the ten richest men in the world, who had called me here.

Norton Clapp had long been a presence in my life. I first heard his name from my childhood classmate Mike O'Rooney. One of the benefits of parochial schools was the wearing of uniforms, which masked all class distinctions. Every girl in the same forest-green jumper, every boy in the same salt-and-pepper pants and forest-green sweater. It was not until the sixth grade, when Mike invited me for a sleepover at his house, that I learned he lived in a mansion. I had never been in any home as large as Mike's. It felt like a castle, with layer after layer of floors and wings that ran off and turned corners. There was even a hidden TV room just for kids in one of the turrets. Mike could only laugh at my stuttering amazement. He explained that his house was nothing compared to the Norton Clapp mansion. As Mike put it, "Compared to his, our house is a *shack*."

The next morning he took me out on the lake in his speedboat, and when he pointed out the Clapp mansion, I could see what he meant. It looked like it had been plucked up from some estate owned by European royalty and carefully set down here amidst the trees of the evergreen forest. He said his dad told him Norton Clapp was the richest man in the world.

The man behind the desk directed me down a wide hallway to a reception table where the president's trim, athletic assistant quickly set me up with a name tag pinned to my navy-blue V-neck sweater. She informed me that the trustees were behind schedule. Sitting on a single chair with its gold plastic cushion, I reviewed my talk. My plan was to give the trustees a sense of the expanding universe by telling them a story of how Hubble might have come to his insight. Making up a story about Hubble but keeping all the science accurate. If I could give the trustees a feel for it, they would recognize Hubble's significance and would want his work featured in a university course.

I had no doubt whatsoever that the Hubble event of the early twentieth century would come to be regarded as one of the milestones in the evolution of human consciousness. One of the dozen or so pivotal events of human history, on a par with Siddhartha Gautama reaching enlightenment under the Bodhi Tree or Shakespeare composing his plays in Southern England, Hubble's experience was comparable in terms of the largeness of impact. Each of these three events changed human understanding of the universe forever. Mt. Wilson was in Southern California, only a two-day drive from the University of Puget Sound. Directly south. In the future, throngs would be processing up Mt. Wilson to touch the Hooker Telescope just as pilgrims today make their way to Rome and Mecca. A two-day drive to be there, to be near the telescope where Hubble had his searing insight, to be at the exact place where evidence of universe's birth first pressed into human consciousness.

THE BREAKTHROUGH event began with Hubble sitting atop Mt. Wilson, peering out at the vast ocean of stars. The light arriving at Mt. Wilson was an actor in this dramatic event. It was the reason Hubble and his colleague Milton Humason worked in the cold of the San Gabriel Mountains year after year. The light provided the justification to build the one-hundred-inch telescope. The light was the reason two tons of fused glass had been melted down to form the core of the telescope's mirror, which required an entire year before it cooled enough to be coated with pure silver that would reflect and concentrate the waves of light traveling those vast distances through the universe to reach Mt. Wilson.

Hubble asked a single question: "Have these light particles traveled millions of years before arriving at Mt. Wilson? Or have they traveled merely several thousand years?" In 1928, the answer was unknown as Hubble sat there with his telescope looking hour after hour at the points of light.

Henrietta Leavitt offered Hubble a deeper way of seeing the light. She had discovered something profound concerning the nature of Cepheid stars, stars that grow brighter for a number of days and then grow dimmer for the same number of days. Leavitt had learned how to calculate the amount of energy the Cepheid star was radiating second by second. By drawing on her method, Hubble was no longer restricted to observing stars as just points of light. He could now observe their fundamental dynamism in terms of their energy production.

An additional element in Hubble's mind was the mathematics of Johannes Kepler. Kepler was the first to publish what came to be called the inverse square law for light's luminosity. The geometry of the universe via the inverse square law reduced the star's enormous radiant energy down to a paucity of photons reaching Hubble's eye there atop Mt. Wilson. To get a feel for Hubble's awareness, then, three aspects needed to be highlighted: the white dots of light in his telescope, the reality of mathematical law, and the memories of his previous viewings.

Hubble's breakthrough began with the Cepheid stars in the direction of the Andromeda nebula. After months of patiently attending to the phenomena of the light from Andromeda, after months of carefully bringing the mathematical physics of Johannes Kepler, Henrietta Leavitt, and others into the process of his apprehension, there came a moment when Hubble pierced the confusion. In my imagination, I saw Hubble's shocked face when he suddenly realized the photons of light he was viewing had been traveling on the order of a million years to reach him. He suddenly knew the star he was watching was far beyond the Milky Way galaxy. Some scientists had speculated that the Milky Way was the whole universe. Others had speculated that there were additional galaxies beyond the Milky Way. But now, here was Hubble, alone in the night, looking at a star in the Andromeda galaxy and experiencing himself as receiving starlight that had traveled a million years or more to reach him. In that moment, he became the first human to know, in a direct and empirical manner, that we live in a vast ocean of galaxies.

To HAVE discovered that there are galaxies in the universe beyond the Milky Way was already a historic achievement. But Hubble's full awakening involved the movement of these galaxies. As before, Hubble's mind was shaped by many thousands of previous achievements, most crucially the insight of Niels Bohr, who discovered that when an element such as hydrogen moves from one quantum state to a lower quantum state, it emits a photon of light. Each chemical element emits photons that carry its own signature. It's as if hydrogen atoms emit photons in C major and G sharp minor keys, whereas carbon atoms emit photons in the D minor and D major keys. Each of the ninety-two elements creates photons with a unique vibrational signature. Once Niels Bohr and his colleagues learned this, they could analyze the vibrations in each sparkling star and know which elements were present there. This was true whether the sparkling light was from the Sun or from a star a million trillion miles away.

In order to gauge Hubble's consciousness as he peered into his telescope, one needs to understand that what for most people would be experienced as "a pattern of white dots in a circle of darkness" was experienced by Hubble as a kind of symphony. As he gazed into the night, each star and each nebula deluged him with notes the chemical elements were playing. Subtle cosmic music flooded him from every direction of the universe.

There is one final feature that brought it all together.

Hubble realized something strange. The galaxies were "singing at different octaves." The galaxies were humming similar songs, but some galaxies hummed two octaves lower than other galaxies, while others hummed the same song ten octaves lower. To say that the songs were sung at a lower frequency is to say that the waves of their music have been stretched out.

But that was exactly what Friedman had shown Einstein when he demonstrated, mathematically, that Einstein's equations spoke of the possibility of a universe expanding in all directions. A universe expanding in all directions will stretch out all the waves of light coming from the galaxies. Hubble was experiencing this directly. He had found himself in the midst of these stretched-out waves as would be the case in an ever-widening universe.

In one silent, thunderous instant, the developing universe surfaced in Hubble's mind. He had knitted together observations and calculations. The songs of galaxies the farthest away were singing in the lowest octaves. He, and he alone, was experiencing this. *Because the universe is expanding.* He must have repeated that phrase over and over. He repeated it over and over because he was torn in two different directions. Out of habit, he perceived change as something that happened to objects in the universe. But the data were whispering a truth radically different. They were suggesting that the universe as a whole was changing. Light from the more distant galaxies was stretched to lower frequencies because galaxies were rapidly expanding away. The experience was breaking apart structures of his mind, transforming his understanding of his placement in the universe.

In my imagined reconstruction of the event, Hubble next positioned the Hooker Telescope so that he could view, at the same time, two different galaxies. One of them ten times as far away as the other. Checking his data, he found that the distant one was expanding away from him ten times as fast. Imagine him locating another galaxy, twenty times as far away. Checking its number, he found it was expanding away twenty times as fast. Another galaxy was found to be expanding fifty times as fast. Another, one hundred times as fast. The concept of number had appeared long ago in human history, and Hubble was now using it to wade into a first understanding of this order in the universe. Georges Lemaître's imaginative theory was verified in Hubble's concrete experience. The data from the trillion galaxies was harmoniously interconnected because at the beginning of time there was a primordial explosion that had sent everything soaring.

In my imagined reconstruction of the event, I would have the stars and galaxies whispering to Hubble their never-to-be-forgotten truth: "Our movement through time and space is an elegant expansion. We have been separating like this for billions of years. Now, for the first time, this is noticed by someone on Earth. You. You can now *think* the universe. You alone know that all of us, all the millions of galaxies you can see and all the billions you cannot see, all of us have been rushing apart since the beginning of time."

They were right to assert that he and he alone experienced this in a flash. It is also true that Georges Lemaître and Henrietta Leavitt were also there in the time-developing structures of his mind. Lemaître and Leavitt and a billion others. The discovery of cosmogenesis was the work of the entire human species with its astounding ability to fold back onto itself. When an infant is born, she has a nine-month-old brain, yes, but she sits inside a culture holding hundreds of thousands of years of thought. By granting humans the power to accumulate knowledge, the universe was bringing forth a planetary mind.

Tom Davis, the dean of the College of Arts and Sciences, poked his head out and apologized for running late. As I shuffled my seven pages of typewriter paper, I marveled over the power of the human imagination. Like an inert lump on the gold plastic chair, I had not moved in half an hour, and yet in that time the whole story of Hubble's epic discovery had danced through my mind. For two hundred thousand years, the light from the origin of the universe had bathed us, without any awareness on our part, but then, via the power of thought, the entire fourteen billion years of creativity came alive in Hubble. The whole story had always been there, showering down on us, but so much had been required to develop the mental space to allow it in. Now that a pathway had been constructed into our awareness, we would be changed forever. That's what I needed to get across. Give them one whiff of that and watch them go wild with wonder. I would make this a moment they would not forget.

27.

Isaac Newton Inside the University of Puget Sound

Twelve or thirteen trustees sat behind a long table in the front of the small room. At the center was Norton Clapp, poised and waiting. He squinted at me like a boxer sizing up a newcomer to the ring. The trustees on either side of him were huddled in groups of twos and threes, talking among themselves. Physically, the scene was very close to what I had subconsciously assumed it would be, a long row of serious, august, seasoned judges. But energetically it was far more subdued than I had imagined. The trustees were clearly fatigued after their three days of meetings. As it was late in the afternoon, there were coffee cups and half-filled water glasses scattered about the table with its white cover. The dean introduced me quickly and took his place at the end of the table. Most of the trustees continued murmuring in their several conversations. Should I wait until they were done? Or was this the way they listened to the presentations? I stood there saying nothing for a full fifteen seconds. Davis nodded at me to go ahead.

I began by telling them that this course would be rooted in contemporary mathematical cosmology, but its focus would be to experience the universe's expansion itself. Before I could say more, the trustee sitting next to Clapp stopped me. His white sleeves rolled up, he had an extraordinarily thin face and the thinnest nose I had ever seen, no wider than a pencil. He started off with a serious mien but there was a touch of lightheartedness in his voice. Glaring at me, he said he didn't know the university had a Home Economics Department. I stared back, at a loss. He picked up a sheaf of papers that were stapled together. He folded them back to a specific page. Presumably this was the document containing the course descriptions each professor had been asked to submit. He bent forward and read the small print and pretended to make a discovery.

"Oh! Cos-*mol*-ogy. Not cosme-*tol*-ogy. For a minute there I thought we were getting into the eyeliner business!"

He chuckled and looked both ways up and down the table for an ally, but his humor landed with a thud. The woman next to him smiled and shook her head. They were putting up with him by now. With a big grin, he threw his hand at me as a signal I should continue. I started over. I began by telling them that when future historians consider the magnitude of our discoveries in contemporary mathematical cosmology, and when they compare these to Copernicus's discovery, they will conclude Copernicus was nothing more than a minor blip in the history of thought. The woman seated next to Dean Davis raised her hand. She took her glasses off to address me.

"Professor Swimme, Connie Sandler. Yes, we've read your write-up, very good. But I need some help. Where will your course be housed?"

I did not understand. It would be one of the three university trustee courses, wouldn't it? I looked over to Dean Davis and gave him a quizzical squint, seeking help.

"I assume the Department of Mathematics?" he said, looking at me.

"Sure," I said.

"So this course will satisfy a math requirement?" Connie asked.

"I guess," I said.

"You don't mention that in your statement."

She looked to the dean again. He saw that I was at sea.

"What Connie is asking is whether the Math Department has signed off on it."

"Wouldn't this be one of the 'university courses'?" I asked.

"Yes," Davis said. "But each is housed in a particular department."

I was stymied. I considered saying that certainly the Department of Mathematics would be happy with it, but that was not actually clear at all. I stood and said nothing. After an awkward moment, the trustee at the far right end of the table took over. His enormous bald head sported a ring of white hair above his ears. He touched the stapled pages on the table in front of him. He seemed restless, straining at the leash.

"I'm not surprised the Math Department doesn't support your course," he said. "It's not a math course. What is it? Help me out here."

Less than five minutes had passed. I had a feathery feeling in my stomach. They seemed to have already made up their minds. They wanted to wrap things up. The black hands of the wall clock showed it was just after five. But I had started late. I assumed I would still be given a full half hour. It was clearly a do-or-die moment. Either I reached down inside, right now, and came up with something undeniably convincing, or in two minutes I would be limping away as an also-ran. Some of the trustees were shuffling papers. I hesitated a moment then made my bold decision.

"If I could quickly show you the equations," I said.

I moved to the whiteboard off to the right where the schedule of the day had been written out. I picked up the cloth from the tray and erased the left half of the board. That would be enough room. In the uneasy silence, I wrote out Einstein's field equations. As soon as I saw the equations on the board I had a better idea. I erased them and wrote down the metric for the de Sitter space. I needed to give them a lightning-fast introduction to non-Euclidean geometry. I needed more room, so I erased the rest of the board. Dean Davis lifted his hand to get my attention, but I positioned my body to keep him out of my peripheral vision.

"We start off with a space of all possible universes, okay? Now, if we put in the dynamics from Einstein's equation, we find a topology that is a stunner, a complete surprise to scientists."

Dean Davis stood up to put an end to it. With a wisp of a smile, perhaps out of compassion for the awkwardness of the moment, he took the blame. He explained that with the very full schedule, they had gone over time. People had plans. They needed to get back to Tacoma. He thanked me for my presentation. Though on some level I understood it was over, I couldn't move. I had a mouthful of ideas that I *knew* would convince them if I could just release them. A few of the trustees were gathering up their belongings. The man with the huge bald head was shuffling quietly behind the backs of the other trustees. Worst of all, Norton Clapp, sitting in the center and wearing a blue knitted cap, was asleep. Or pretending to be asleep. Chairman of the board, arms folded on his chest, eyes shut.

The meeting adjourned; Dean Davis came over. Davis was the archetypal hire of President Phibbs. Even better than Ivy League, Davis had earned his PhD in mathematics from Cambridge University. With that one move, Phibbs had placed the University of Puget Sound's Mathematics Department in the direct lineage of Stephen Hawking, and further back, Isaac Newton himself.

"Too bad." He shook his head.

"Is there anything I can do?" I asked.

"Sorry, Brian."

28.

Seattle Space Needle

The trustees said their goodbyes to each other in small groups. Clapp, now with eyes wide open, interrupted his conversation and lifted an index finger to me. It was a gesture that asked me to wait a minute. Or was it? It was not completely clear what he meant so I stood there pondering what to do. Which would be worse, to continue standing there stupidly, or to leave and risk suggesting to the chairman of the board that I thought I was too important to wait? I fretted over this until my indecision decided for me.

With the last of the trustees departed, Clapp sorted through his papers to put them in order. He wore a blue suit that looked twenty years old. Knowing how much power he wielded and how much money he controlled tightened my chest. Wearing his blue knit hat even though the room was now quite warm, and with facial features blurry from the retreat's demands, Clapp locked his papers in a black leather briefcase then looked at me. He pulled a folded sheet of paper from his coat pocket, tapped it against the padding of his left hand.

"President Phibbs gave me three names. Yours was one of them." He put the paper back in his coat pocket. His eyes narrowed.

"President Phibbs's dream is to make the University of Puget Sound the Harvard of the West. What's your sense of this?"

He steamed ahead before I could answer.

"You're happy here, or so I'm led to believe."

"I *love* Puget Sound! I couldn't be happier."

He turned away. Without looking back at me he lifted his right hand and indicated with his index finger that I should follow him. He stopped before the floor-to-ceiling glass windows that formed the western wall of the conference room. I stood beside him. It was odd. As he stepped close, I thought I saw a fine white scar running from his forehead down to his cheekbone, as if he had been in a sword fight and had narrowly avoided having his eye put out. Maybe he had lost an eye. Maybe he was using a glass replacement.

He gestured with his hand.

"What's that?"

It was February twilight. We stared through the reflected light to a grassy hill that sloped down into a wooded area at the shore of Hood Canal.

"Hood Canal," I said.

He shook his head.

"Bigger."

"The Puget Sound?"

We stared hard for a moment at the hill and the trees and the water. He pulled out a ring of metal keys and used one of them to dislodge something from his ear.

"The world's most massive gold mine. IBM at a penny a share. It's so obvious it's hard to see."

He scratched his scalp with his free hand underneath his blue knit cap.

"Do you know who built the Seattle Space Needle?"

He looked at me. I had no idea who built the Space Needle.

"You?"

"Can you guess why I built it? Why I went ahead with the idea even after the city of Seattle dropped it?"

I shook my head. I struggled to take in the fact that I was standing next to the man who had built the Space Needle. Even though on one level there is nothing impossible about that fact—after all, everything has to be built by someone—on another level I had been lulled over the years into thinking of the Seattle Space Needle in mythic terms. It had simply appeared one day. When Grandma Trudy invited my sisters and me to go and behold it, Mom and Dad made clear to us the gigantic expense this was for our grandmother. She was a licensed practical nurse at Seattle General Hospital and lived way out in West Seattle. It took over an hour and two bus rides to get to work in the morning. After changing bedpans until eight at night, she was back out in the rain to wait for the buses to take her home again, often to endure the assaults of an alcoholic husband. From this hardscrabble routine, she squeezed her daily expenses until she could obey the universal summons and take her three grandchildren to pay homage to a new world colossus.

As a child of eleven, I was stunned into silence when I first saw it from the distance, dominating the Seattle skyline. Then I found myself standing beneath it, leaning back to see it whole. Three sets of curving parallel lines swooped up to the sky. The girders themselves were a creamy yellow, as if they were the bones of some giant reptile. But these were not dead bones from the Cretaceous. Seen from below, the vast wheel these girders supported seemed to come alive. The several dozen steel struts packed closely together were a creature's gills that would, any moment now, commence breathing. A surreal marine animal, far above, ready to lift off and return to its cosmic ocean at the right signal.

I had never thought of it as an architectural structure a human could make. I was being forced to take it in. The Seattle Space Needle had actually come into existence because of this guy standing next to me wearing his blue knit cap. So ordinary looking and yet capable of making stupendous events happen, things as impressive as the Egyptian pyramids.

I was talking with Zeus.

As we stared at the Salish Sea, Zeus sketched his plans for making the University of Puget Sound the intellectual leader of a new civilization in the Pacific Northwest.

"Every great power has the same three components." With his thumb and fingers, he ticked them off. "Natural resources, technology, military supremacy. We have all three. Weyerhaeuser owns more nature than any company on Earth. Boeing is the world's largest producer of aircraft, military and civilian. In technology, the computers developed here will change the business world. We are heading into an economic explosion bigger than Paris and London and New York put together.

"I understood next to nothing from your presentation." He held up his palm to stop me from speaking. "Nor do I wish to. I have no interest in the past. I came to this event because I want to build the future. I'm looking for leaders."

He nodded curtly and severed our conversation. He marched off with torso rigid, a commanding general. Maybe he had a military background. On the far side of the conference room, a custodian vacuumed the dark red carpet. The grinding sound rose as he pushed the chrome-plated machine forward, dropped down when he drew it back. He worked the carpet between the aluminum legs of the whiteboard I had smudged. The vacuum's racket was a dreary finish to the afternoon, especially in contrast to the promise the whiteboard offered when I first entered. I wanted to avoid the awkwardness of encountering Clapp again, so as he made his way toward the lobby, I went in the opposite direction, exited through a back door, and took a path through the trees to the parking lot.

29.

The Olympic Peninsula

I raced through the night on the two-lane road that curved through the black and silent fir trees of the Olympic Peninsula. The forest crowded both sides of the road, so with every curve my headlights tore gashes of light into the darkness. Glancing at the white needle of the speedometer and seeing it jittering just past sixty, I pulled my foot off the accelerator. I was disturbed. It was obvious I had failed with my presentation. As I drove and remembered my words and the responses from the trustees, I grew even more embarrassed. But beneath the turmoil, there was a mounting excitement. Norton Clapp had laid out his plan, his way into the future. *Norton Clapp.* And it included me. Didn't it? Maybe it did. That's how he started out. Was he challenging me? Had I ruined things with my presentation?

I crossed the suspension bridge over a marsh at the center of which a dark river, swollen by the recent rains, flowed swiftly out of the forest toward the unseen Hood Canal, the trees wet and dripping from the rains. I had been convinced my course was a breakthrough to an original relationship with the universe. Without hesitation, they had tossed it on the trash heap. What did that matter anyway? Wasn't the Space Needle more important than a single course? Wasn't changing the world with Clapp more important than anything I could do in a classroom? Shel's news had so shocked me, I clammed up on the ride back from his new house. Only when we crossed the Narrows Bridge did I find the energy to say something real. Hoping to change his mind, I explained my question about his childhood. I had been certain that he and I were the same and that one way or another a fascination with the stars would have been part of his early years. He said nothing when I finished. In a weird switch of topics, he told me the gist of a short story by Issac Bashevis Singer. The opening scene presents two twentysomething revolutionaries. Their story morphs into a scene sixty years later when they bump into each other on a Miami sidewalk. Now as overweight eighty-year-olds, they laugh at their identical pink, flowery shorts, gigantic enough to stretch over their large bellies. I didn't see the connection.

"You want me to shed a tear over some loss of innocence," Shel had said. Ramping off the bridge, climbing up to Sixth Avenue, he had offered me a cigarette. I passed. Keeping one eye on the road, he sucked in the flame of his Bic lighter and released a stream of smoke that hit the windshield and spread across the dashboard. "Look, drop the self-righteousness. Maybe you're right, maybe we shouldn't have a military. Maybe Russia will just lay down its ICBMs. Or maybe they'll blow us to bits. I don't pretend to know the ultimate truth of things. But I'll tell you what's *real*. The cost of college tuition. And I've got twin boys."

He picked a flake of tobacco from his tongue. Almost as an afterthought, he suggested I join his working group at Boeing. Get in on the ground floor. "It's so new it doesn't have a name yet. The working handle is 'the Black Box,'" he said. "Who cares about a name when you've got bottomless federal funding."

I could be part of that. Keep my position at Puget Sound. Clapp might like having me connected to Boeing. One step would lead to another. Was this my route to a house on a cliff overlooking the Puget Sound? In spite of Dad's neurosis around money, handling it as if it were radioactive? And

Mom's Christianity, certain money was the source of all evil? Shel and Clapp and a whole lot of other people thought otherwise. It was my chance to leave childhood behind.

As I sailed through the night, it started to rain again, just a drizzle at first, then big drops that needed the windshield wipers. I flicked the radio on and rolled down the window. The high-pitched whir of the tires against the asphalt. The *swish swish swish* of the wipers. High up, far above everything else, the moans of the wind that whipped through the dark branches of evergreen trees. All that consternation I had created. Over nothing. My tenure decision was only two years away. A bone-deep shiver passed through me. I was safe now. I knew what I had to do.

30.

Rocky Point, West Tacoma

The very next morning, boiling with excitement, I met with Dolores Maro. I assumed she would be thrilled by this offer from Clapp to transform the University of Puget Sound into the major intellectual center of the Pacific Northwest. My plan was for the two of us to meet with Clapp. The first step was getting her on board. She and I started at the grotto and strolled down the narrow path between the Puget Sound and the railroad tracks. To the west, across a mile of sea, was the forest of the Olympic Peninsula. To the east, beyond the railroad tracks, was the shallow lagoon where gray herons hunted for fish among the pickleweed and cattails. I waited as long as I could. When I couldn't hold back any longer, I blurted out the vision of building a new civilization centered on Seattle.

I finished and waited for her to speak, but she said nothing. She looked at me with a half smile.

"You don't find this exciting, do you?" I said.

"I'm not so sure," she said. "No. That's not true. In fact, I don't find this the least bit exciting. Just the opposite."

Her words threw me off. I made another run at it.

"I know what you're thinking, but Clapp is *for* the university. He's the same as we are on this. We could create something powerful."

After a pause, she spoke. She chose her words carefully.

"Please don't misunderstand me. You are a fresh and enthusiastic mathematician. I regularly hear praise from your students."

Dolores brushed her thinning gray hair back with one hand. She pulled out a flap of her shirt to wipe the lens of her glasses.

"How do I say this politely? I've tried several times already. Brian, you have to imagine we're living in Germany when the Nazis took over. They've asked you to be a dean in a university and you're all atwitter. Don't you see what would happen? Whatever came out of your mouth would be twisted into their demented ideology. Here's the worst part. Your own words wouldn't require that much twisting. Anyone trapped inside a sick culture becomes pathological just to survive.

"It's not your fault, Brian. You're a victim of the modern industrial university in its decadent phase. I admire what I perceive in you as the truth quest. You do have the potential of developing a rough-and-ready philosophical outlook. But to do so will require your becoming keenly aware of the way in which your education has drained you of your humanity."

My throat closed up tight. I could hardly speak. I barked out a question. "Drained me of my humanity?"

"Not just you," Dolores said. "The university has become a toxic dump, and that includes everyone, professors, administrators, and most especially trustees."

"So you are not immune either?" I spat out the words.

"No one likes hearing he is an undeveloped human being. You will forgive my blunt speech. I don't know how else to say it. At the very least, your present emotional turmoil will enable you to appreciate the difficulty of Socrates's 'know thyself.' Ignorance is easier. Fantasizing that we are healthy and whole is easier. Remaining deluded about the ways we've internalized industrial society's deformations is easier."

I was locked up with fury.

Whenever she had lamented the prevailing forms of consciousness in modern society and especially when she bludgeoned the corporations and their control of university life, I had always assumed she regarded the two of us as different from the rest. That we had committed ourselves to a more impressive life than that. That we had chosen the path that soared above all the money-grubbing and status-seeking compulsions she saw all around us. Clearly not. A seagull cried out high in the sky, three long plaintive notes, adjusted its wings slightly, and swooped down to land on one of the large, angular, dark brown rocks that kept the winter seas from washing up over the railroad tracks. It preened its beak on its gray feathers as if sharpening a knife. The foul odor of rotting fish wafted up from the shore.

"Please give me the courtesy of telling me why you have concluded I am a subhuman," I said.

Dolores marched on in silence, the sea on our left, the lagoon down a short hill to our right. She stopped and turned on me.

"Your error is to think you can change the shrunken world in which we live. You can't. Your idea for a course that would enlighten the masses failed utterly. But instead of sobering up, you fall into Clapp's vision for the future, a man who made his fortune destroying the old-growth forests of the Pacific Northwest. If only I could show you how lost in illusions you are.

"Plato lived near a sea very much like the Puget Sound. He had some noteworthy observations, like his claim that we live in a cave making it impossible to see things as they are. He was right. You look at the Thompson Hall of Science and imagine an English Gothic temple filled with Pythagorean scientists pursuing ultimate truth. Those are the shadows in your cave.

"If you desire clear vision, rip off the ivy and paint it the black and brown camouflage of military aircraft. *That's* what takes place in your revered Thompson Hall. The Defense Department owns American science.

And while you're at it, ditch your childish Catholic belief in a benevolent God. Seeing reality as it is disappoints us, yes. But delusion is worse."

WE WALKED on in silence. I was a witch's brew of hurt and confusion. How had my closest colleague come to think so little of me? What was I doing? What was I missing?

31.

The Long View Down 15th Street

When I returned home I called Ron VanEnkevort, chair of the Mathematics Department, and set up a meeting. I didn't want to put it off even one day. The main thing was it had to happen soon. The stress that had been building up inside me was breaking down my health. I had been deteriorating for weeks. My legs, even in warm weather, felt cold as icicles. They twitched spasmodically. I had no idea what was taking place. It felt as if worms were under my skin eating my calves, which caused them to squirm and shake. I bought woolen long johns and bulky winter pants. At St. Vincent de Paul's I found an old fireman's coat, its cloth thick enough to deflect falling walls. But even on the warmest days, I shivered inside my heap of clothing.

We met on the fifth floor of Thompson Hall, in the room where two to three dozen professors from the natural sciences and mathematics gathered every weekday at ten for their morning coffee. But as it was Saturday, the room was empty. Ron was casually dressed in jeans and a red plaid shirt.

Though he glanced at my strange clothes, he avoided addressing my condition. I didn't know what to say, so I began with an apology for applying for the trustees' course, which he had been against. I knew that once I got that off my chest my health would improve.

He received my apologies without any rancor for my stubbornness. He said he was delighted I was embracing the department in this new way. It was what he had been hoping for. He would set up a night course that I would teach out at McChord Air Force Base the next semester. He explained how this was a huge opportunity in that the air force would pay the tuition of any airmen who wanted some training as they transitioned to civilian life. If the course went well, if they were satisfied with what they got from me, we would be guaranteed a continuous stream of new students.

As he spoke about all this, my mind began to fog up. Someone was hanging a large gauze blanket between me and the world. Not all at once. Slowly. I noticed that I lost some of his words. Ron had begun talking about one of the most important features of having a professorship at the University of Puget Sound, the generous retirement program. He wanted to help me set up my financial future. He noticed that I had elected an annuity, when experience showed it made much better sense at my age to take the risk and invest in equities. He said he understood how a young professor might want to proceed with great caution, but if I took his advice he was virtually certain my fund would grow quickly and I could retire in my fifties.

As he spoke, I drifted like a cloud even though I did not move from my chair. One of the standard bits of advice for avoiding seasickness on an ocean liner is to place your focus on the horizon; as if intuitively following this marine procedure to deal with my foggy mind, I moved my eyes up a fraction of an inch from Ron's animated face to gaze out the window behind him. I stared straight down 15th Street, which headed west to the Puget Sound and beyond that to the great ocean that connected the coffee room to the round planet.

It came without words. I could feel it. The history of the universe and the history of life had eventuated in this moment here at this wooden table, staring down the pathway taken through so many potencies, so many intuitions and fears. Naturally, inevitably, there would be an infinite number of dreams that rose and died. It made no sense to dwell on those. I felt addressed by the journey itself. This was my destiny, sitting here with Ron VanEnkevort, his face a mask of concern.

"All that matters are the students," he said.

He spoke of the parents who put out significant funds to bring their children to the University of Puget Sound, that we as professors needed to remember them. Our appointments were not for research but for students. When he confessed that his primary motivation was to help the students have families of their own, admiration swelled in my heart, for Ron personally, for all the dedicated professors. He closed our meeting by saying that, contrary to how it might appear at times, he did support my intellectual interests. He offered me leadership of the Math Club, which met Friday afternoons and which he had been running for several years. He said this could be a wonderful setting for my ideas combining science and philosophy.

We shook hands with tight grips. His smile expressed both relief and happiness. Besides my gratitude for his help, I too was deeply relieved. We had made things right by getting to a new and simple understanding of my responsibilities. Surely my health problems would soon be a distant memory.

32.

The Observatory of Puget Sound

As I finished up a long day, putting my books on the shelves in my office, I set off across the quad wearing my black hat and buttoning my long fireman's coat up to my chin to hold off the cold wind. Someone called my name, trotting toward me on the pathway from Jones Hall. Tall, dark hair, awkward as she approached, Monica DeRaspe had taken my calculus course the previous semester. She had a question for me.

"It's not from your course." She giggled, nervous. "It feels like everything is infinite. That there is an infinite amount that can be learned. About anything. All this truth surrounds things. It doesn't matter what. I can see it, this infinite richness everywhere. It's like it's calling me forward? Do you think that's weird?" she said.

"I don't know what you mean."

"Is it because I'm unique?"

"Unique how?" I said.

"I don't know. Maybe God is speaking to me?"

I STOOD there defeated. After an awkward silence, she mumbled an apology and walked off toward Jones Hall. Not wanting to give her the impression I was following her, I headed back toward my office, banging through the oak doors of Thompson, taking stairs two at a time to the fourth floor. The passageway that led to the tower was in the middle of the building. One flight up was the coffee room where Ron and I had met, dark and vacant. I scrambled up the stairs that led to the sixth floor, then out the small door into the cold night air, up the steep ladder to the open roof.

The towering evergreen trees of the quad cut jagged shapes of black in the gray sky as they swayed under the strong wind. I searched for the stars, but the cloud cover blocked them out. In perches like this, ancient astronomers contemplated the night sky from twilight to dawn. Tycho Brahe on an island in Europe. The Anasazi in the deserts of North America. Though I could not see them, I knew many billions of galaxies lived beyond those clouds. Each galaxy with a hundred billion stars. Tonight I could see only a dozen stars with the dark clouds blown in from the ocean.

I KNEW what was going on. Her question upset me because her question was my question. Infinity. We think we've taken it into account with our mathematical symbols, with our infinite summations, with our infinitesimal distance. It openly mocks these attempts. Even the simplest form, the ratio of a circle's circumference to its diameter, is a transcendental number beyond our reach. We calculate to a thousand decimal places and we feel good about our approximation but it's an approximation, not the number. Calculate to sextillion decimals. Still not there. Even if the approximation is serviceable for our engineering projects, there is a sense that we haven't really made any progress, for no matter how far down we've drilled with our calculations, an infinity of decimals remain, unseen, laughing at our Sisyphean failure.

The origin of the universe is an infinity that expands into the known universe. We don't know the meaning of a universe coming from infinity. It's the same with our knowledge of the end, with the universe expanding into an infinite future. That's what draws us on. Why couldn't I just say that to her? When I speak of the earliest moments of the universe, I am only too happy to say the expansion rate is evidence that the universe knows how to construct stars and galaxies. Well, if the universe in the form of primordial plasma knew where it was headed, why shouldn't the contemporary universe know where *it* is headed?

The fir trees were suddenly bathed in light, as if someone had snapped their photograph with a flash camera. I looked behind me and saw the next column of lightning, then the crack of thunder from the first hit. The winds kicked up even higher and made the swinging branches of the Douglas firs into a giant instrument of sound. Another flash, this one a double branch of lightning, the thunder tight and loud, the storm coming directly at me. I heard it first. It sounded like heavy chains dragged across sharp gravel rocks, then the wall of water hit and plastered my hair against my skull. I was one of the highest bodies on campus other than the trees, a passive lightning rod, but I held on to the brick wall, a captain taking his whaling ship into the source of the thunder and lightning. I uttered words, a kind of prayer directed at the storm: *Tell me what to do.*

33.

A Multimillionaire in the Basement of 1508 North Cedar

My confidence that meeting with Ron would bring back my health proved to be hollow. Without warning, my pulse would shoot up to 250 beats a minute, each beat an electric jolt in my chest. I had to hide the chest pains from Denise. The first time they hit, she demanded we go to the emergency room at Sacred Heart Hospital on North K Street. By the time we got there, it had begun to ease up, but the staff insisted on putting me through a battery of tests anyway. I sat on the table naked but for the green cotton blouse they provided. Doctor Cohen stood in front of me with bowed head studying the EKG.

"You might have had a heart attack. Then again, you might not have." His voice was soft and matter-of-fact, but when he looked directly at me, he seemed irritated. "Are you under any additional stress?" My life was collapsing. I was waking up at night streaming with sweat. I did not fully know what was happening, but I knew if I admitted this out loud I would break down.

I shook my head.

"Why are you so bound up?"

AFTER THE hospital visit, I used alcohol to deal with any heart pain, freeze-drying my system with a quart of vodka. No matter what time of day it might hit. This plan worked well in that as soon as I dulled the pain, the heart seemed to calm down on its own. The hard part was when it kicked in and no alcohol was available. This was a challenge because there was no rhyme or reason as to when it would start. I had just driven home after teaching and was sitting in the car and found Denise and Thomas Ian playing together in the front yard. She wore pink slacks and a white blouse covered with pink roses. She waved excitedly when she saw me. They were about to try something dramatic. She steadied the trike by holding the handlebars with one hand and the back of the seat with the other. As Thomas studied the lawn's two-foot drop before him, I watched from the parked car.

With his stiff little arms, he managed to hold the front wheel straight, and when he and his trike glided to a stop right at the edge of the sidewalk, he seemed bewildered that it had come to an end. He immediately looked up at me as if hoping I would help him make sense of it, or perhaps applaud. I left the car there at the curb, got out, and shut the door. I knew by the sound it had not shut completely. It was a defect in the door that I knew only too well. Looking around behind me I could see that, yes, the door was ajar as usual. Half an inch off from being shut. I reopened the door and slammed it hard with both my hands on the window.

That's when it started. From something as innocuous as slamming the car door. The pain was one thing. The discouragement of having it come back again and again was worse. I passed wife and son without saying a word, clomped up the steps, heart pounding in pain. In the refrigerator was a six-pack of Rainier beer. I pulled a bottle out, had a better idea, jammed it back next to its brothers. In the cabinet above the stove, behind some cans of corn, was the 750-milliliter bottle of vodka. I pulled out the clear pitcher of orange juice from the refrigerator and poured some into a half-full glass of vodka. I drank it down while standing there at the sink. Denise followed me into the kitchen, peering at me questioningly.

I picked up a yellow sheet on the counter next to the sink. At the top were the letters "J & J Electric." In the bottom right corner, in blue ink, "$39.95."

"The furnace is broken," she said.

"Perfect," I said. I poured myself a second drink of orange juice and vodka. "I've been feeling at loose ends lately. This is very good news."

"I'm taking care of it." Denise pulled the receipt from my hands. "They're installing a new one on Tuesday. Ours is fifty years old and leaking fumes. There's a crack in the wall."

"That's what they always say." I took a big drink from my glass. "We're not taking any money from your father, Denise. No way. We'll turn off the furnace and wear heavy coats if we have to."

"I'm not asking Dad," she said. "I can handle this."

"How? We can barely pay our bills."

She bit her bottom lip. That gesture with the look in her eyes was all I needed to figure it out. I finished my drink and poured another three fingers of vodka into the glass, topped this off with a touch of orange juice. The warmth of the alcohol was already easing me out of the pain in my chest.

"Let me have a look at it," I said.

THE DOOR to the tiny basement was next to the stove. I held my orange juice and vodka in one hand and used the other to navigate the nearly vertical wooden steps down. There was only a six-foot clearance so I had to crouch down to avoid hitting my head on the joists. I pulled the cord to the dangling light bulb and bent down to assess this contraption coated with grime and sporting pipes and wires and switches. It had been lurking down here for many years waiting for its opportunity. For some unknown reason it had chosen me as its target and was now threatening to blow up the house in a final act of savage revenge. I sat down next to it and took another long drink from my glass, then got on all fours. The basement floor was gritty with mice droppings and ancient dust. I sniffed the air; the oil fumes seemed the same as ever. But wasn't carbon monoxide odorless? I peered into the darkness until I found a small metal switch down at the base. I flipped this in both directions. I had watched Dad labor over our own furnace when I was a child, so I knew this was the right move. Nothing happened. This was the extent of my knowledge. I flicked the metal switch a few more times, hoping for a technological miracle.

In this deepening gloom, the truth was becoming clear. I had drawn the same diagram in dynamical systems so many times. I was in the middle of a branch point. Confronted with a question at its most basic level. The essence of the situation had little to do with this hideously malfunctioning furnace. I drained my orange juice and vodka. Everything was all right. I rolled onto my back, the alcohol blurring my disgust for lying on top of the black pellets from the mice. I had wanted to contemplate the mysteries of the universe by pure thought, and this is what it led to.

My father-in-law had suggested I become an actuarial scientist. A senior vice president at Rainier Bank in Seattle, he had set up a meeting with one of his vice presidents so I could learn about the job. I was then a second-year graduate student in mathematics with a desire to understand the dynamics of cosmic evolution and with little interest in business, but I wanted to convince him that I was a good son-in-law, that I would provide for his daughter and his grandchildren.

We were on the thirty-fifth floor of the Rainier Tower on 5th Avenue. It was the corner office of one Herman Ozeal, a kind, elderly, baldheaded actuary who was explaining his work. I was arrogant. With two years of graduate school mathematics in me, what could he possibly know that I didn't? He brought out chart after chart of data and spoke quickly and efficiently in the language of mathematical statistics concerning acceptable risks in the complex systems of the nuclear, oil, and transportation industries. It was quickly becoming clear that he would be just as successful teaching a small rodent. Decades of concentrated study had led him to a tiny branch of highly specialized knowledge. My confidence concerning my command of mathematics changed into anxiety that he would catch my eye and realize I was at sea. The whole enterprise had no interest for me, but at the same time it was a path one could take to get beyond financial anxieties. Starting pay was more than my parents' combined salaries as high school teachers. After a series of tests, with corresponding raises for those who passed, the money really started to roll in. As he clipped through the escalating remunerations of my future salary, I did the multiplications and realized that sitting opposite me was someone who had ridden these numbers into the zone of the multimillionaires. I wondered what kind of pull my father-in-law must have to get a senior officer like this to spend his valuable time explaining his profession to me.

It ended poorly. Mr. Ozeal suddenly put his charts away. Perhaps he intuited my lack of interest in calculating risk tables. He asked me what I was studying. I answered him in a voice that was perhaps too loud. I told him I was interested in the universe and how it came to be. He stared with dull eyes. I had assumed that as soon as I mentioned this he would deluge me with questions about the birth of the universe. But no. Nothing. I found it hard to believe that he really didn't care. How could he wake up each morning and not wonder about his existence in this universe?

He closed the three-ring binder with its brown leather cover. To finish answering his question, I told him I was hoping to carry out my investigations as a mathematician in a university on the West Coast.

"Professors make very little," he said. "Senior administrators do all right, I guess. Hognes, the president of U-Dub, has a small condo in Hawaii. You're young. When you get to be my age, you'll learn. Money beats squalor."

I had nothing but pity for him when he trotted out this lame excuse for a life purpose. I considered it a waste of time to listen to the advice of an accountant burrowed into the thirty-fifth floor of a bank building. But my self-image of moral superiority had certainly not helped me navigate the real world. Take a good look at yourself. He's up in a tower and you're in a basement with a dead furnace. You can't provide warmth for your family.

Who's the lame one here? The worst part was having to ask my father-in-law to buy us a furnace. Not directly, no. By using the money from his wedding gift. That was her plan, surely. We had put it in a savings account for our son's college tuition. She was forced to use that money because of my failure to come through on money matters.

OUR BASKETBALL coach at Santa Clara University, Dick Garibaldi, had lectured me on financial irresponsibility in front of the whole team. Even though this took place nine years ago, the memory penetrated through my foggy inebriation as I lay on my back on the basement floor. We had just lost to the University of Nevada at Las Vegas in an ugly game filled with technical fouls and shoving matches. Ranked fourth in the nation, we had been the heavy favorites to win the game, so they were delirious with their stunning upset. At the final buzzer, the UNLV fans swarmed the court. We had to push our way to our locker room. One of the men antagonized Garibaldi by walking backward in front of him, smacking him on the chest with both hands, saying, "How'd you like *that*?" When he did it a third time, Garibaldi grabbed the man by his tie and hit him with the fist of his other hand, splitting the skin to the skull. With blood running down the man's cheek, Garibaldi pulled him nose-to-nose and screamed: "How'd you like *that*?"

The next morning, we were waiting to board the bus, holding our red game bags. Garibaldi marched up frowning. Without any warning, he used his concentrated energy to drill into me. It was not my play on the court that bothered him. I hadn't gotten into the game at all.

"Swimme, how much did your breakfast cost?"

I had no idea what it cost. But now I understood why he had requested to see the individual bills after the waitress collected them. I had ordered three orange juices, so it was much higher than what the other players had cost.

"Five fifty!" He jabbed his hand at me twice, fingers splayed, stopping inches from my nose. That was just the warm-up. His fury fastened onto my conversations in the locker room. From now on, whenever I opened my mouth, I had better make sure I was reviewing the offense, the defense, or the inbound plays. For everything else, I was to keep my mouth *shut*.

It was a sunny Nevada morning. Most of the players took a seat for themselves so they could stretch out and get some sleep, but Bruce Bochte and I sat together near the back of the bus. Coach's speech had frightened me into silence. I resolved to stare out the window the entire trip. The resolution lasted half an hour. Once we were out of the city and flying up Highway 15, I had to tell Bruce an amazing fact I had learned in Professor Barker's class. It was the asymmetry of the universe. In hushed tones, I conveyed the main points. In the first instants of the universe's existence, there was so much heat that an enormous amount of matter and antimatter was drawn forth from the quantum vacuum. But it all happened with a strange, asymmetrical twist. For every billion plus one particles of regular matter, there were a billion particles of antimatter. Which meant that when the universe cooled a bit, a colossal collapse took place as the matter and antimatter collided and exploded into light. Only a tiny bit of matter remained. One particle of matter for every billion pairs of annihilating particles. Which meant that the present universe, now, was one-billionth of what was originally here.

Dr. Barker dwelt on the strangeness of the process. Creation and obliteration happened at the same time, but because of that asymmetry, a tiny remnant escaped. Why did the universe create a billion particles in the beginning to have a single particle survive? Why start with a billion times more than you need and then have almost all of it disappear? Why not just start with the particles you will use to make things? He drew a parallel with the life process. Over the history of Earth, something like three billion different species have come forth. But the vast, vast majority of them have gone extinct. Life constructs billions of species that flutter around and enjoy existence for a moment or two and are then destroyed. But the destruction is not complete. A tiny remnant survives into the future.

I finished by asking Bruce a question: "What could be stranger than a cosmic process that annihilates a billion species of life? Obliterating all but a remnant? Isn't that bizarre? Destroying all but a tiny fraction of what has come forth?"

I had been whispering to make sure Coach, who sat in the first seat of the bus, couldn't hear me. When I stopped, Jolly Spight, one of our teammates, sat up. He had been slouched down in the seat in front of us. Tangled hair, wild eyes, he gave his one-sentence evaluation.

"Swimme, that's some heavy shit."

As I stared at the floorboards of our house, I felt better. The memory of massive death cheered me up. This cosmic dynamic clearly was at work in my own life. My dreams were swirling down the drain, yes, but there had to be a creative remnant somewhere. It would show up. Out of nowhere. I would ride that creativity into the future. I thought to myself, *That is your saving insight.* I heaved a long sigh of relief, the air passing through my nostrils and bringing to mind the pleasant scent of vodka. That needed to be remembered too. This saving insight was coming from someone lying on mice droppings after having downed three screwdrivers.

34.

The Piano at the Winthrop Hotel

The day following John Lennon's murder, Denise and I were at dinner in Tacoma's Old City Hall built on the bluff above the tideflats. It had stormed all day. The rain washed over the glass windows and blurred Commencement Bay. In the middle of the meal, I went to the men's room and there on the wall of the stall was written, "John Lennon, December 8, 1980. The day the music died." When I saw that sentence and imagined the hand of the person scrawling it on the gray metal of the stall, it broke in on me, the reality of this moment. When I returned to our table, the tone of our dinner changed. Denise and I were thrust into reflections on our fierce love of that music, how much it mattered to us. Now he was actually dead. He and all of that time in our lives were gone.

It was the second time music had died in my life, the first happening in eighth grade. I loved playing the piano all through early childhood, learning from Sr. Alberta Louise. Her pupils practiced down in the basement of the convent at St. Frances Cabrini, working each year on pieces we would perform at the archdiocesan competitions held at Holy Names Academy in Seattle. All of this came to an end when I entered high school. Bellarmine Prep had no music. At first I continued playing on my own, purchasing song sheets from Wilson's Music in Lakewood Center. But as the months went by, I played less and less until I finally jettisoned it altogether. I forced it so far out of my mind that it wasn't until the senior prom at the Winthrop Hotel that I was startled into remembering. And this only because of a fluke.

Everything about the prom was glamorous: the cars lined up outside of the hotel, my friends emerging in tuxedos and formal gowns, hurrying up the red carpet to get out of the rain, entering the vestibule and finding the older crowd dressed up as well, chaperones and hotel staff, lined up to the top of the elegant marble staircase that led to the Crystal Ballroom. As soon as we got through the massive entrance doors, I rushed my date through a network of alcoves and hallways, which led to still more rooms, and in one of these, so distant the crowd became a faint rumble, we came upon a grand piano set up next to the window overlooking Tacoma's harbor. Without thinking, I sat down and began playing. It was the first time in years. A simplified arrangement of a song by Johann Sebastian Bach, my favorite. She didn't know I played the piano. I had never mentioned it to her or to anyone else in high school.

As Denise and I reminisced, it suddenly dawned on me that the Winthrop Hotel was only a couple of blocks away, just up the hill. Why not go there right now? Maybe the piano was still there. I could try to play on it. I had forgotten all the pieces I once knew, but I could play a simple chord progression. Just to feel that elegant ballroom filled with beautiful sounds one more time. It was raining slightly when we got outside, but it was not the rain but the excitement that made us run south on Commerce Street. It was all so familiar, the lights off the pavement, the swoosh of the cars in the dark. Charged with these memories, I entered a sense of time travel. We were running up the hill at South 9th to cross the physical distance between Commerce and Broadway and the temporal gap between spring 1968 and winter 1980.

When we reached Broadway, there were no lights from any grand hotel.

We walked to where I thought the entrance should be and learned the truth. Though the building was the same, the Winthrop Hotel was no more. It was the same entrance, but grimy now, layers of grit on the brown brick walls that no one bothered removing. The double glass doors that had been added to the entranceway were locked. Inside, on a stand, was a sign with somewhat faded gray lettering that read, WINTHROP SENIOR CONDOMINIUMS. I struggled to believe the meaning those words announced. At first I thought we should check the next street, but the truth was right in front of us. The Winthrop Hotel had been converted into senior citizens' apartments. On some level I had thought, as we ran up the hill in the dark, I would see the bright shining faces of my friends once again. I would join their youthful shouts of joy and their awkward humor at finding themselves in formal dress wear.

In that instant, standing there in the rain with Denise, the previous night's dream came back. It had awakened me with its violence but had vanished until that moment when we looked at the now run-down Winthrop Hotel. I gave Denise the essence. "I'm holding a carton of eggs, which is glued shut. I know the eggs are going bad. A mob of men attack me from the shadows, and I slash them with a long chef's knife. Then you and I are running down Orchard Street toward my childhood home. Our tunics are soaked with blood but the Sun is shining and soon we are running in luminous white clothes."

Denise was clutching her arms against the damp cold.

"Your precious egg," she said.

"What precious egg?"

"The one you're always guarding. Your creativity. That's what the dream is about. Your creative energy is turning rancid."

She put her arms around my neck. We were both shivering.

"We should go," she said.

35.

Epiphany at St. Patrick's

At Mass, March 15, 1981, eight o'clock Sunday morning. St. Patrick's church, a block down from Aquinas Academy, the sister school to Bellarmine Prep in North Tacoma. Out of the blue, it began. There was no reason for it. And not the slightest hint of forewarning. It came in the middle of Fr. Burne's sermon. He had been wondering if there were other planets with humanlike life; if there were, he speculated that theologians would begin to ask some radical questions about the Christian faith. Would Jesus have to come down to each of these planets? To sacrifice himself again and again in order to save these humanoids from sin?

No vodka at hand, I sat in the pew and breathed slowly, hoping this might dissipate the torment. I prayed for help. I prayed that the suffering would stop, that it would lighten, that it would lift, that it would dissolve altogether and leave me alone. After ten minutes of praying with no change, I switched to calculating how long I would have to wait until I could use vodka to extinguish the fiery electric throbbing. Ten minutes until the end of Mass. Another four to walk to our car. Five to seven for the drive up 12th Street, over to Yakima, left on Cedar. One minute for the stairs, another minute to put the blender on the counter, fifteen seconds for the frozen orange juice to be pureed. No, not frozen orange juice, gulping that down would give me a headache. Something warmer. Kahlua. No, that was gone. Just water, then. Forget the blender. Equal parts vodka and water to anesthetize the pain.

Denise and I left the dark interior of St. Patrick's and emerged into the sunny morning on the sidewalk of North J Street with the church bells pealing brightly. Conversation and laughter from groups of parishioners gathered around the three stainless steel coffee makers on the Formica-topped tables. Donuts in pink cardboard boxes. The pavement was scattered with the white blossoms from the cherry trees, and a light breeze dropped a new wave upon the children running about with cries of glee. The evergreen trees that lined the street were bursting with sap. It was a scintillating spring morning.

A YOUNG woman in tennis shoes, wool slacks, and a white shirt sporting a large green apple approached me. Her beaming face was framed by a surrounding halo of beautiful red hair. She held a donut in one hand. I knew that I knew her, but her name would not come. Only when she spoke did I realize it was Oona Fitzgerald, one of my favorite students. Her excitement was so great she stuttered to get her words out.

"Dr. Swimme, I-I got the fellowship!" she said.

I was cringing. It was awkward, seeing a student at Mass. I disliked any mixing of my professional and private lives, and the awkwardness was made worse by my ignorance concerning what she was talking about. Perhaps I had written her a letter of recommendation, but the throbbing in my chest shredded my memory. She bubbled over with her news. When word of the award came, she and her roommates purchased a bottle of champagne, her first taste. When she called her parents about the details of the offer, they were sure she was joking. She was the first member of her family to attend college. She'd grown up in a trailer court in Sumner where her parents still lived. When the voice on the phone told her the size of her annual stipend as a graduate student, she asked them to repeat the amount.

"I don't deserve it." She shook her head gravely. "I really don't." I offered some stilted conversation to celebrate her achievement. I wanted to affirm her hard work and her brilliant mind. Her eyes narrowed. "I couldn't have done it without you."

I hardly knew what I was saying. I was trying to get through the moment so that she would go her way and I would be released to the vodka.

"What university?" I asked.

"MIT or Stanford. I can't decide!"

"Your fellowship is not from a university?"

"That's the coolest part! I can take it wherever I want. Isn't that nifty!" Her eyes gleamed. She took another bite of the donut, still watching me. The same joyous face, the same wide-open curiosity as when she asked about the meaning of life in the special relativity course.

"Who is the award from?" I asked.

"Dow Chemical," she said, her mouth full of donut.

I NODDED and held on to my frozen smile. Without any sound, something broke. She waved a shy goodbye and skipped away, shaking her red curls, celebrating the spring day. This was what Dolores wanted me to see. It was right in front of my eyes. I taught science to children so they could poison the planet. All those years of study, all those decades of learning, all of that used to turn Oona into a tool. Not out of malice, only ignorance, pathetically ignorant of how my actions in the world revealed the true meaning of my life. I had been so certain she was wrong. Dolores, the cynical one. I, the sunny person of faith. She had predicted it all. Dow Chemical had hooked its talons into Oona.

Standing there on the sidewalk with apple blossoms under my feet, immobilized, I stared at the flesh on the back of my hand. I had no idea what I should do. Sit down and hope the pain would go away? Ask Denise to take me to the hospital? There was nothing unusual about my hand. Everything was the same; the same black hairs, the same softly throbbing veins. Jean-Paul Sartre's characters were revolted by the absurdity, the nauseating facticity, of little hairs growing out of skin. I was not revolted. I was spellbound. The Sun's light was warming my skin. I had never once stopped a moment to reflect on this miraculous event. Human skin had the power to transform the physical photons of a star into something nonphysical, my experience of warmth.

These science facts I'd known since high school, but I'd never lived in them. Detonations in the Sun a hundred million miles away were striking my skin with its black hairs. The Sun, in my skin, transformed into experience. Thermonuclear explosions in the core of a star jangled the electrons of my skin and became—by the powers of my nervous system—feelings of warmth.

But Earth itself was forged out of elements created by a previous star, a star that had ended its existence in a supernova explosion that sent its elemental creations screaming into the Milky Way where they condensed into our solar system. That previous star, unnamed and unseen, had constructed the elements composing my skin. I felt warmth from one star, and the gift of existence from another star. I was not separate from this process, "having" this experience. I was in between. This awareness was in between two stars. Though separated by trillions of miles and billions of years, the creative forces of these two stars—one that warmed me, another that built me—were fused in an awareness that I was part of a gigantic process, a gigantic, creative process that was me.

36.

"The New Cosmic Story"

Denise convinced me to go to Whidbey Island to be sure of my decision. When I arrived, the office was empty but Fritz had left a page of instructions on the round oak table in the farmhouse. It was here, a year before, that I had questioned Fr. Lank O'Connor about key authors rethinking the world. In his welcome note, Fritz indicated that I was to stay in Bag End, a small cabin at the southern edge of Chinook. Before I made the trek in the dark, I chose some reading material from the makeshift library consisting of four wooden shelves with an assortment of books and pamphlets.

Chinook published its own journal, and as I leafed through several issues one title caught my attention, "The New Cosmic Story." I did not know the author, Thomas Berry. As I read the opening paragraph, it felt as if I were reading something I myself had written. Except that the understanding behind the words was far more comprehensive than my own. More profound. Here was a person who seemed to know all the important things I had discovered and much more. He spoke of the music of the universe. He spoke of cosmic creativity. He spoke of the universe haunting us. I read straight through as I stood in the library of the farmhouse. At the climax, he explained what he meant by a new cosmic story: *My own feeling is that we need a complete change to a creation mystique. Adaptation of the cosmological myth born of scientific understanding and openness to the forces governing its unfolding evolution constitute our greatest challenges.*

My mind was combustible matter that his words set ablaze. *Creation mystique.* I had never heard that phrase before. Even more penetrating were *cosmological myth* and *new cosmic story.* I did not have any sure hold on what these might mean. That didn't matter. What I knew beyond doubt was that my destiny was tangled up in those words.

IT WAS still black night when I awoke the next morning. Using the electric pot I had brought from home, I made myself a cup of coffee and sat in the wooden chair at the small desk. The dark fir planks of the walls had two windows without glass so that the night air wafted in through the lacy curtains. From the desk drawer I pulled out a sheaf of white vellum paper. One side was blank and the other filled with the notes and cleft symbols for the Gregorian chant. Someone had cut the sheets of music into quarters and stowed them there in the desk should any guest need writing paper. I positioned the first sheet on the wooden table, stared at the page, and took sips from my coffee. In the stillness of the morning, I waited for the world to wake up. The forest of scrub oak and alder, though still shrouded in night, would soon light up with the first rays of dawn.

It felt as if I were writing my obituary. I was giddy. The deed itself of making the decision was now in the past. I took a huge lungful of air, released it, and the words spilled onto the page. I thanked President Phibbs for his vision of interdisciplinary education. I thanked the Mathematics Department for giving me a chance. I wrote at lightning speed, putting language to whatever surfaced. I thanked the Jesuits who had trekked to

Tacoma and constructed Bellarmine Prep. I thanked Dad for taking me fishing on Chambers Creek and throughout the Salish Sea, for waking me up before dawn to hunt deer in the Cascade Range and duck on the lakes of Eastern Washington.

As I wrote, a new thought broke in, that a specific cosmology had been transmitted to me before I could think, before I could speak, before I could know my name. And even now, a grown man, my foundational cosmology came from those early years. Dad, growing up on the banks of Fraser River, was a culmination of one tiny branch from a hundred thousand years of human evolution, and Mom, from her place near Alki Point in West Seattle, was another culmination, another tiny endpoint in a long, branching development. Like all humans, Mom and Dad had absorbed a system of assumptions about reality from their ancestral relations. They married these together and handed them on to their own children—my two sisters, Leslie and Nettie, and me. That's where my cosmology began. Not with Lemaître's theories. It began when Mom nursed me, when she enveloped us children in ten thousand acts of love, none of which we would remember but that would shape our sensibilities throughout our lives.

I thanked Leslie and Nettie for our journey together in that ineffable world of infancy and toddlerhood. And for teaching me profound truths. My earliest memory was Leslie making me laugh, again and again, conveying her intuition that I was important, that I, her baby brother, mattered. And when Nettie appeared, she latched onto another dimension of our parents' cosmology. She taught us, in her subconscious way, the metaphysical hunch that kindness is what matters more than anything else. My sisters put in place the foundations of my cosmic sense of things.

When I returned home, Denise and I talked into the early morning about what I had found in Thomas Berry's "The New Cosmic Story." I needed to take on the challenge Berry had articulated. I needed to study, needed to learn so much in order to do this right. I could no longer stay at the University of Puget Sound. Later that morning, I phoned Bruce Bochte, who was down in California with the Mariners playing the Oakland Athletics baseball team. Still vibrating with excitement, I ran through the main ideas of Berry's article. He instantly understood the monumental sense of it. We made plans to get together when he returned, but half an hour later he called back to tell me he had spoken with Linda, his wife, and they both thought now was the time to set up our center. That I should take a year and do the necessary research in order to tell the new story, and that he and Linda would cover all the costs.

37.

Goodbye to Dolores

When I called Dolores to tell her the news, my hand was shaking like a teenager asking a secret crush for a date. Though we hadn't spoken in months, I had thought of her continually, rehearsing again and again our hard conversation and how I had missed the truth she tried to convey. Her voice over the phone lines was aloof, more formal, but the main emotion her tone expressed was mild surprise at hearing from me. When she asked why I was calling, I told her I would rather not tell her over the phone. I said I was hoping I could come by and see her, promising it would not take long.

When I arrived, Dolores did not step out of the house for our usual peripatetic walk. She greeted me cooly and invited me in. She set me up on the reddish-gold armchair that looked out on the garden with its rows of newly planted vegetables behind which was a small orchard of apple, apricot, and peach trees. Dolores returned from the kitchen with freshly brewed coffee in white cups on white saucers. She set these on the glass-topped wooden table in front of the couch, which ran orthogonal to the picture window.

When I told her I was resigning from my position at the university, her face fell. She stared at the coffee cup on the table. In my trauma of trying to decide what to do about my professorship, to leave or stay, I never considered whether Dolores would miss me. She asked if I realized I was throwing away my career as a mathematician. I shook my head, telling her I was certain I'd keep up the mathematics on my own. She didn't reply but her eyes registered her skepticism. When she asked if I had thought ahead a few years to fully understand the consequences of this move, I was startled. Though we hadn't been in touch with one another, I was certain she would praise me for finding the courage to leave. Hadn't that been the point of so many of our conversations over how the corporations and the military owned the universities? I told her none of that mattered. It was over, I was leaving. She leaned back on the sofa, taking it in. A silence of defeat replaced her heated interrogation. I knew this would be the last time we spoke, so I needed to tell her the truth.

"I was so naïve," I said. "I cringe when I think of how I argued against you. It took me so long to see." I waited for her to look up, but she kept her eyes averted. I went ahead and said it anyway. "You breathed Pythagorean fire into me. I'll never be the same."

"I was hard on you," she said.

"You had to be. I was so pigheaded."

She met my eyes briefly. When she looked down again, I noticed the delicate spidery lines in her eyelids, the tracings of old age. She held the skin around her lips in a tight bunch.

"You have avoided the prisons of economism and militarism," she said gently. "And you are finding your way out of scientism. You should know these temptations and others, even more virulent, will continue. It is not going to be easy. Constant vigilance will be required. The shrunken industrial university is too small for your vision, but do not slide into the fantasy of thinking there is an ideal institution out there somewhere. No matter where you go, forces that run industrial society will try to impose a smaller version on you. Fight back. To avoid discouragement, keep your high ideals before you every day. There is no substitute."

I HAD to leave. I was already late to pick up Thomas Ian from his playgroup. I wanted to launch forth into one of our long conversations, but that was no longer possible. To break it off, I stood up abruptly. Dolores walked with me without speaking. She moved out to the middle of the road in order to guide me past the voluminous laurel hedges. Once more, the thought that I might not ever see her again entered my mind. When she returned to the driveway, I gave her my two-finger salute. She pressed her lips back against her teeth in what was her smile. Just as I was about to accelerate away, she lifted her hand up and walked quickly toward me. I rolled down the window and she leaned her elbows against the car door.

"I apologize for my poor response to your announcement of your daring move. It irks me that in that moment I could only think of myself. Please understand that my initial lack of enthusiasm does not express my true feelings. I couldn't be more proud. But the truth is, I will miss you. I do not have anyone else to talk to, not compared to our conversations. Having you here has been as close to having a son as I have ever experienced."

"Dolores . . ."

"One last thing. Now listen to me. When things get hard, and they will, you will sometimes wonder if you should have stayed here. When that happens, I want you to remember what I am about to tell you. You will find a guide. You have a path ahead of you, a path I cannot travel. But you can. You have the necessary spiritual resilience."

"Spiritual resilience . . . Which means?"

"The ability to leap off cliffs."

SHE PRESSED her lips together and backed away. Her eyes had already brightened. Standing firmly by the side of the road, she radiated the serenity she had layered into herself through decades of study and reflection. She was a true descendent of the Greeks in their clear-eyed acceptance of the bitter aspects of existence, of death in all its forms. I sped away, unable to think. I had never met anyone with such wisdom. When I stopped at the red light, I slammed the white steering wheel with both hands to stem the tears spilling down my cheeks.

38.

Stars in the Salish Sea

The next week, at night, I slipped letters under the doors of the president, the dean, and the chair of the Department of Mathematics. I had already officially resigned, but I wanted to leave each of them with a final expression of gratitude. Especially President Phil Phibbs, who had made the extraordinary offer of creating a new department for me. If such an offer had come at any moment in the previous three years, I would have leapt at it. But by the time he spoke to me, my imagination had already left Tacoma. I did not know how to explain any of this to him. I simply thanked him for his generous spirit and his vision of what a university could be.

When I got home, Denise and I and Thomas Ian left the city driving west on 21st Street, south on Proctor Avenue to 6th Avenue, where we headed west again. Cars were sparse. We passed Goofy Goose, the hamburger joint, then Reman Hall, the detention center for adolescents. We glided quietly down the long hill that became the Narrows Bridge, which carried us across to the Olympic Peninsula.

At one point out on Highway 16, we heard muffled explosions somewhere off in the distance, possibly from Fort Lewis. I knew the sounds from my childhood in Lakewood but was surprised to learn that they traveled across the waters of Puget Sound all the way here. The U.S. Army staged war games in the vast forests owned by the military, using mortar rounds that were fake, of course, but just as loud as real ordnance.

WHEN DENISE heard the explosion, she looked over and said, "I guess Phil just opened your letter." The absurdity of her joke broke through my anxieties and I laughed and laughed at the silly image. But it was hollow laughter. What came to me was a story of a group of zoologists who rescued a dying monk seal, nursing it back to health. When it was deemed ready to survive on its own, it was lowered into the ocean in the release cage, darting from one side to the other with eyes opened wide as the wire doors were lifted. Instead of rushing away, it lingered in its prison cell and had to be pushed out. Just so did I now find myself. Constraints were dropping away. Phil was no longer my president. I had no job, no place in society. Dad's ambition for his only son, that too was gone. After we crossed the Narrows Bridge, I groaned when I noticed the fuel gage just above empty. There was probably enough to make it, but to be sure I started gliding down all the hills to save gas. On the long downward slope to the Purdy Spit, I shifted to neutral and we coasted through the night in silence, the only car on Highway 16. Near the bottom, I let the clutch out and the engine gave a loud crack, like a backfire, and died. We had enough momentum to turn left off the highway and cruise over the short bridge that separated the mainland from the narrow spit of land. I pulled onto the shoulder of the road and tried to start it, but it wouldn't catch. After several more attempts, the battery started winding down. The machine was dead. Tacoma was miles behind us. Key Center was just about as far ahead. It would take hours to walk either direction. I got out and put the hood up, bracing it open with the long metal rod on one side.

The Purdy Spit separated two dark bodies of water. The high winter tides had left a tangle of seaweed and driftwood on the sand that was black in the night. I undid the plastic safety belt of Thomas's car seat, bundled him up inside a large blanket, and carried him in my arms down to the beach. Denise found a smooth log for us. We'd wait for help. Surely someone would happen by.

Sitting there in the quiet, we watched the water become black glass, utterly still. It was the slack tide when all the salt water that had rushed in from the Pacific Ocean through the Strait of Juan de Fuca swelled the Salish Sea to its highest mark and held itself there, as if basking in a moment of triumph before Sun and Moon drew it back to the ocean.

The dark clouds had thinned enough to show the stars. Orion was suspended low in the sky as if frozen in ice. One of its stars was predicted to explode, and I focused my attention on its light arriving here after a five-hundred-year flight. Had Betelgeuse already blown up? Was this the last light from a star that no longer existed? I looked down at the salt water and studied the shining stars quivering on the surface above the unseen kelp. It was similar to the experience in the amphitheater. The stars flickered below me in the Salish Sea. They twinkled above me in the night sky. They dwelt inside me in the watery world of awareness. As I stared down at them, the stars moved with the tiniest ripple, perhaps from an unfelt breeze, or from the flow of electricity in my nervous system. I felt I could stay on this spit forever. I had wanted to be a professor of mathematics and to discover the synthesis between general relativity and quantum field theory. Had I given that up for something so flimsy as an experience?

I turned to Denise.

"There is no way I could have done this without you," I said. "I've put you through so much. You deserve better and I'll get there, I promise you." Denise leaned her cheek on the top of Thomas's head and smiled. Her eyes shone.

"Your dad is so worried. I've never seen him cry.

"He had such high hopes," I said. "I know I've disappointed him deeply."

"It's more like you're following in his footsteps. He left Fountain as a teenager to join the marines and fight in a war. You're tame in comparison."

THOMAS SQUIRMED in her arms. Rocking him gently back and forth, she tucked a strand of hair behind her ear. The cold was penetrating our clothing and I knew it was time to go back and wait in our wreck, but when I offered to take our son, Denise held him close.

"JUST A little longer," she said. "Who knows when we'll be here again?"

IN THE trunk of the Simca were half a dozen long, red cylinders that looked like sticks of dynamite. Dad always insisted on them. In all the years of our childhood, we had never once had an occasion to use the flares on any of our family car trips. Once I pleaded with him to start one just so I could see it in action, which he did for me in our garage at Leona Way, scraping the end of the cylinder on the concrete floor to ignite it. Completely disappointing. I had assumed it was going to be something like a Roman candle that would light up the sky, not a tiny sparkle at the end of the red stick.

I LEANED the sandy driftwood together to make a small cone and trained the red flare on the thinnest wood chips. It took a minute to catch fire. The flame was barely intense enough. We huddled under my fireman's coat with the mist blowing in from the Pacific Ocean and the cloud cover spreading apart to reveal a clear night sky. She fell asleep in my arms, still holding Thomas Ian in hers. Without her strong presence, my anxieties came crawling back. She had been ecstatic when the letter came from the University of Puget Sound. She would be able to return to the Seattle area where she had grown up. She could be near her parents and her siblings and friends. Our sons could grow up with grandparents and aunts and uncles and cousins. I had thrown that away. I had made so many mistakes. Was this another one? The biggest of all?

PART TWO

*

Stars and Galaxies

39.

The Professional Fool

It was in grad school at the University of Oregon when I first became interested in the idea that a full understanding of the universe requires more than the objective, rational thought of mathematics. I loved the mathematics; I had devoted more of my life to its study than to anything else. But could mathematics alone convey the fullness of the universe? Or was something more required?

This something more, also known as "other ways of knowing"—the nonconceptual, conative, intuitive, heart-centered, mystical—refers to knowledge that comes not from an objective analysis of the thing but from a communion experience with the thing, the prime illustration being the knowledge one gains when falling in love. I was wondering if these more "subjective" ways of knowing had any place alongside mathematical science. Was it meaningful to speak of a heart-centered knowing of the origin of the universe? Would it be as real as the equations that described the universe's birth and development? Could intuitive knowledge be synthesized with mathematical knowledge to form an even deeper understanding? And if so, could I myself attain such a synthesis?

In exploring these ideas, one line from a book by the theologian Matthew Fox caught my attention: "What does it matter to me if Mary gave birth to God thirteen hundred years ago if I do not give birth to God every day." The words came from his translation of the fourteenth-century German theologian Meister Eckhart. I hardly knew how to understand this. In my Catholic upbringing, the phrase, *Mary, the Mother of God*, appeared here and there but I had never given the words any serious attention. By the time I was an adult, all theological thoughts had been pushed to the furthest corner of my mind. For me, *Mary, Mother of God* referred simply to belief in the incarnation. That and nothing more. And yet here was this strange idea from Eckhart about giving birth to the ultimate of ultimates each ordinary morning.

I would have tossed the book aside if it were not for the explication that Fox provided. He brought out subtleties in Eckhart's thought that gave a glimpse of the universe rather than an explication of theological doctrine. Over time, I began to see a formal congruence between Eckhart's intutions concerning emergence and the fundamental dynamism of quantum field theory. Eckhart spoke of a human being giving birth to "God" in every moment; quantum physicists spoke of the quantum fields giving birth to virtual particles every moment. Should these considerations be sequestered in their separate disciplines of thought? Should scientists have nothing to do with theologians?

I wrote to Fox with the suggestion that since science and theology are both central to civilization, a newly discovered resonance between them might signal the appearance of the next era of Western civilization. I asked him bluntly: "Do you agree that we are in the beginning stages of building a global civilization?" Fox responded with handwritten letters containing triple exclamation points. I took the next step and asked him if we should invite others into this investigation. He wrote back and said I needed to meet Ken Feit since Feit was about to perform in the Pacific Northwest.

Fox described him as a "professional fool." This was disappointing. I had assumed Fox would recommend someone living in a Trappist monastery, hidden in a forest, on the top of a snow-covered mountain where monks like Thomas Merton contemplated the deep mysteries of time and existence. Or if not that, then perhaps a cutting-edge scientist like David Bohm or a profound philosopher like Alfred North Whitehead, not someone who called himself a professional fool.

To meet Feit, I drove to a private home in the town of Mukilteo, twenty miles north of Seattle. Feit's audience consisted of four people gathered in a living room with two stuffed chairs, a gray couch, and a dark fireplace. He sat on the dark blue shag carpet in a half-lotus posture, wearing a jean jacket with dozens of patches. He began by asking us for stories of our foolishness. I was depressed and embarrassed by the scene. After each of us confessed some ludicrous act from the past, Feit acted out one of his "playlets," short dramatic pieces designed to convey his message.

When my turn came, I told the group I had just resigned from what had been my dream job as a professor in a university. As quick as a hard rubber ball bouncing off a concrete wall, Feit said, "A lot of people are succeeding at things that aren't worth doing in the first place." He poured several drops of oil in a spoon, dropped a kernel of corn in it, and held it over a burning candle. We sat in silence and watched the kernel. When it popped up into the air, he turned his head in an arc, at glacial speed, and locked eyes with me. He seemed to be saying that I, like this kernel of corn, was about to experience something wonderful.

A week later, he was dead. I found a note Denise had left for me on the floor. I needed to return an urgent call. When I saw the 312 area code, I knew it was Matthew Fox. Sitting at my grandmother's pine table, I listened to Matt's somber voice tell the tragic story of how Feit had died, how he had fallen asleep at the wheel, how this resulted from the way he punished himself as he crisscrossed the country performing his skits in celebration of what he called our divine foolishness. He was always sleep-deprived, sometimes falling into a slumber in the midst of his own workshops. Matt's voice grew angry when he recounted the final moment. As Ken sat in the wreckage his car had become, he was asked by his panicked traveling companion what she should do. He replied, "Please be quiet. I'm dying."

But there was something else. Matt had a question for me. It was the reason he called. It was the last thing in the world I was expecting.

AFTER I hung up, I drove down to Commencement Bay and found it filled with miniature whitecaps. North of the restaurants, no buildings, only shoreline. Parking on the side of the road, I climbed down to the jumble of brown limestone boulders and lay down on one. The terrible news sank in as seawater lapped against the stone. Cumulus clouds the size of huge weightless castles drifted above. My last words to Feit haunted me. I told him I had to get back to Tacoma, that I had an appointment. The truth was that my restlessness made it difficult for me to sit in any one place for too long. That's all it was, restlessness, but I had to pretend otherwise. I didn't want him to think that I wasn't enjoying his playlets. I should have said that the image of the kernel popping up from the spoon was working me over, which was true. I had absorbed much from his performance. To take in more would be water spilling out of a filled glass. He stood and hugged me goodbye, and I was surprised we were at eye level, each of us six feet, five inches tall.

This similarity of bodies was remarked upon by Matt on the phone. For him, it explained why I appeared in his dream, which began with a figure of Ken Feit then morphed into me. Matt pointed out that Feit and I shared a common educational background, which was Jesuit. But there was yet another connection. He explained that Ken was driving cross-country in order to teach a seminar for Matt's institute at Mundelein College. All of which meant that Ken died in the middle of a beeline from my West Coast world to Chicago, a straight shot from Brian to Matt. This gave his dream a spiritual significance involving both of us: Ken had died; Ken had appeared in his dream; Ken had morphed into me; Ken was en route to Matt's institute. Which led to Matt's weighty question.

"Will you come to Chicago and teach Ken's course?"

DENISE WAS in the kitchen, chopping celery for a potato salad. I blurted out the news that someone I had just met had died. Nothing like this had happened to me before. I thought it better not to tell her how interested I was in teaching a course at Matthew Fox's institute. When I asked Matt if there was a set curriculum that Ken was going to use, he laughed and said I could teach whatever I wanted. Denise stopped and listened to the account of Ken's death, but she returned to her celery with added vigor at the mention of going to Chicago. When I started speculating about which authors I'd use for my course, she stopped.

"*You want to go to Chicago*?" She pronounced *Chicago* the way most people would say, "To a cess pool?" With a groan, she returned to her vigorous chopping. "And what about discussing it first?"

"I thought we were."

"Wait . . ." She looked at me. "Some guy had a dream about you, and on that basis, you want to cancel MIT and Boston? Because of this guy's *dream*? Did I already mention this is someone you don't *know*? Can you see how bizarre that is?"

"When you put it that way . . ."

"And this course you'll teach? You're substituting for someone who calls himself a . . . a *what*?"

"A professional fool."

WE HAD already made plans to move to Boston. With Bruce's support, I could apply for unpaid postdoc positions. Carl Sagan, the world's most famous scientist in 1981, sent me a cheery, typed reply inviting me to join his research team at Cornell. In addition to Sagan, I contacted Gian-Carlo Rota, a professor of mathematics at the Massachusetts Institute of Technology, and Vladimir Arnold, at Moscow State University, one of the principals of the Kolmogorov-Arnold-Moser theory I was investigating. They too agreed to my request that I come and study with them for a year.

When Denise and I sorted through the possibilities—Ithaca with

Sagan, Moscow with Arnold, Boston with Rota—her favorite by far was Boston. In her mind, it was one of America's most interesting cities with all its history. She was especially excited about the chance for our son to take in the arts and museums of Boston. As opposed to that, Ithaca was a big question mark, and Moscow was plagued with a string of negatives—a country militarized against America and a track record of long and bitter winters. On top of every other challenge, she was five months pregnant and quailed at the thought of negotiating childbirth in Russia without any knowledge of the language. For myself, I felt it was heroic of her to agree to leave the security of a university position in Tacoma and charge off into what I hoped was a spiritual adventure but which could be an exhibition of stupidity. The least I could do is go with her choice of Boston.

Things had fallen into place quickly. Professor Rota helped us secure low-income housing in Boston. We had our three plane tickets from United Airlines tacked to the corkboard in the kitchen. Now I was throwing that away and asking her to move to Chicago. I had never before made a life decision based on a dream, so I was in agreement with Denise. This was bizarre.

"You're a *mathematician*," she said. "Not a . . . not a whatever you said."

"It's an opportunity."

"Isn't MIT a better opportunity?"

Until her question, I hadn't actually compared the two places. MIT was one of the greatest centers of science on the planet. Did I really want to throw that over to go someplace no one had ever heard of, Mundelein College, a women's college known only for its arts program? I might be the only scientist on the faculty.

"It is strange, but when I think of going to MIT it feels ordinary," I said. "It would be nothing more than camouflage, a place to hide while I worked on the new cosmic story."

Fearing I had said too much, I stopped and waited as she chopped.

"It's just one year," I added. "I'll make it up to you. I promise."

SHE CLOSED her eyes. I knew enough to remain quiet. After an extended silence, she scraped the celery pieces into the brown wooden bowl her mother had given us as a wedding present. I felt a softening, but this might be wishful thinking on my part. She seemed disappointed and angry. But when she looked at me, there was both exasperation and warmth in her eyes.

"*Thank you*, Honey!" I said.

She leaned back and addressed the ceiling.

"Do you have any idea what a *handful* you are?"

"You are the absolutely perfect wife for me."

"Who else could put up with you?"

40.

Matthew Fox in Chicago

The 1981–82 winter in Chicago proved to be the coldest on historic record. When the winds whipped between the skyscrapers, the temperature dropped to eighty degrees below zero. Children were not allowed to go outside for anything longer than a trip from the front door to the bus or car. Television newscasters warned their listeners not to wear earrings because earlobes were freezing as solid as ice before the innocent victim noticed anything unusual taking place. When our landlady, Mrs. McDennebick, saw that Denise was pregnant, she took her coat off and insisted Denise have it to protect the baby. The extreme cold froze the mounds of snow already piled up on both sides of every avenue and in the parking lots of grocery stores and government buildings. Hatchets, pickaxes, and blow torches could not cut through the ice quickly enough to liberate imprisoned cars. It was going to be a long, cold, dark winter.

We rented an apartment in a brick three-story walk-up on Wayne Avenue close to Mundelein College, which housed Matt's institute. It was an immense change. Suddenly I was in a small Catholic college in conversation not with scientists or philosophers but a Dominican theologian, Matthew Fox.

As soon as Denise, Thomas, and I arrived in Chicago, I began working with Matt on a short book exploring this notion that humanity was in the midst of creating a new civilization. Twice a week we would work over our current draft. When the edits filled up the margins, we'd type it up anew. Our thesis was that Western civilization was spiraling downward because two of its pillars—science and theology—needed to be reimagined. This bold claim was not original to us but came from a stream of work following the thinking of Oswald Spengler earlier in the century. My part in our essay was straightforward in that scientists already knew the three-hundred-year reign of Newton's classical physics had ended. The assertion that science needed to reinvent itself in the wake of quantum physics and relativity theories was commonplace. Matt's declaration was both more radical and more controversial. He maintained that the religions of the Western world, especially Christianity, had been captured by the idea that a human's ultimate purpose is to be redeemed out of a fallen world. This fixation on escape had resulted in modern theology's slide into irrelevance, most notably among the highly educated and the young. His proposal was that Western Christianity needed to drop its obsession with getting redeemed out of the world and return to an earlier theology such as that of twelfth-century Hildegard of Bingen, who held that the universe was not "fallen" but the primary manifestation of divine magnificence.

Matt could draw forth an endless stream of treasures from the intellectual and spiritual giants of the West's Middle Ages—doctors of the church such as Thomas Aquinas and Theresa of Avila; forgotten or underappreciated women theologians, including Mechthild of Magdeburg and Julian of Norwich; and radical mystics like Meister Eckhart. We developed our own process for approaching these thinkers. Matt would indicate how their work pertained to spiritual development in the twentieth century; I would follow by drawing parallels between their theological statements concerning the universe and the discoveries of contemporary science. Matt summarized our procedure by saying that by combining premodern spirituality with postmodern science, we were giving birth to a new world soul.

One afternoon in the fall of 1981, my happiness working with Matt exploded out of me. Our notes were spread out before us on the round oak table. I told him in a loud voice that he and I were giving birth to a new cosmological myth of the type Thomas Berry had called for. We were actually doing it.

"He would love knowing this!" I said. "We should tell him it's happening right here on the banks of Lake Michigan!"

Matt smiled.

"So invite him over."

"Seriously?" I said. The thought of actually meeting Thomas Berry had never once occurred to me. I imagined him in a realm beyond reach.

"Call him up. Ask him."

"You've *talked* with him?" I said.

"No, but he's probably got a phone. Well, then again, maybe not. He's a bit of a hermit. The rumor is, he has a stack of unpublished manuscripts stuffed under his bed."

41.

Hildegard of Bingen

On February 11, 1982, Matt and I huddled together outside Mundelein College studying the evening traffic on Michigan Avenue. We were waiting for Thomas Berry to arrive by taxi from Chicago's O'Hare Airport. We stepped from side to side to keep our feet from freezing, both of us heavily bundled to ward off the chill. I wore a fleece vest under my fireman's coat. Matt was wrapped up in a blue jacket with a hood lined in dark brown fur. I peered into every passing taxi to see if it carried Thomas Berry.

Neither of us had met him or seen him in person, but I did manage to track down a picture of him on a conference brochure. His was the only photograph without a smile. He was looking off to the side with what seemed to be an irritated scowl. Irritation often includes resentment, but I saw none of that in his face. His negative feelings came from a deeper place. He was much gloomier. This was a photograph of someone beset with foreboding.

Matt had managed to come up with his phone number and left it to me to contact him and ask if he might come to Mundelein for some lectures.

I studied his picture carefully as a way to prepare myself for the call. As excited as I was about this possibility, I couldn't dial his number. Our apartment was on the third floor with a pantry off the kitchen in the back, which became my office. I would resolve to call him, would even get to my desk and pull the phone over in front of me. Then I would decide I needed a better opening. Maybe I should start off with a confession of how deeply moved I was by his article "The New Cosmic Story." Or maybe I should I tell him something about my own work? Maybe the best approach would be to just jump into our request that he speak to our grad students?

My difficulty came from the way in which his words had penetrated so deeply into me. I had trouble believing a human being living a normal life in New York City had written them. My unconscious assumption was that such words could come only from the deepest stratum. Maybe something similar happened to people who knew Moses or Mohammed. The depth of their visions convinced their followers that the words came from a mountaintop—or a cave—some place where God spoke face-to-face.

There was an even more frightening possibility. What if the opposite proved to be the case? What if he turned out to be narcissistic and career-driven? What if he ended up regarding my phone call as an irritating waste of time? What if the only sentiment he sent through the telephone wires was a desire that our phone call end? My destiny would be squashed out of existence as quickly as it had emerged.

WHEN I finally dialed the number, the voice that answered at the other end of the line was both gentle and open. The possibility that I was actually on the phone with him unnerved me. I asked if I was speaking with Thomas Berry. When he said yes, I asked him how I should address him. That was not my carefully thought-out opening line about how much his work meant to me. I don't know why I didn't just say that. But as he was a monk in the Passionist contemplative community within the Roman Catholic Church, perhaps there was a special way to address him. I didn't know if his group was called "Brothers" or "Fathers." Or something else? I asked him, "Do you prefer Dr. Berry?" Before he could reply I jabbered something else. When my nervousness dropped enough to let him speak, he said, "Feel free to call me anything you're comfortable with." Which led to my saying, "What do your friends call you?" But as soon as I heard the words, my body writhed. Was I claiming to be one of his friends now? Is that what I just said?

Before he could answer, I rushed to tell him we'd love to have him come for several days and deliver some lectures to our students, that we would fly him out to Chicago, that he would receive an honorarium, meals, and a place to stay. I was awkward with all of this, never having set up a lecture series. After I read the logistical details from my sheet of paper, I launched once again into a new topic. This rude behavior stemmed from my fear that he would say no to our petition. Maybe he was too busy? Too uninterested? Too old to travel? I leapt away from our conversation and asked him point-blank if I might be able to read some of his essays. He had gathered together some unpublished essays into mimeographed volumes. Maybe he sold them! As soon I remembered that money might be involved, a new fear kicked in. It came from my experience at the University of Puget Sound when I would contact publishers to ask if I could examine their new texts in mathematics and physics for possible use in my courses. Some publishers, to prevent fraud, required a letter written on stationery with a university's letterhead. Maybe that's what he wanted?

"I'm sure I could get some stationery tomorrow," I said. "I would be happy to send a check as well. What would you need in order to send me the volumes?"

There was a short pause.

"Just your address," he said.

THE YELLOW taxi pulled up to the curb and Thomas Berry stepped out wearing a brown corduroy sports coat with leather patches at the elbows and a thin black tie. He had no overcoat. He wore no hat. A light snow had begun to fall, and when I saw the flakes land on his thin gray hair and melt, I offered him my knitted wool cap, which he refused. He was unconcerned about the cold. Once again, I felt misled by the photograph of him. He was so at ease. The contrast between the somber face on the conference brochure and his physical presence was startling. As the cabbie pulled his satchel out of the trunk and Thomas thanked him for the ride, I searched for the word to capture what it was like to stand next to him. He radiated kindness, but there was something else. As I waited for the right moment to introduce Matt and myself, he stood looking up at the sky through the lightly falling snow.

"Chicago has a spaciousness . . . quite distinct from New York . . . a lovely sense of openness." His voice had a lilt. When he smiled I knew what the word was. He gave off the aura of celebration. As if he were in the middle of a vast cosmic party that I wanted to join with all my heart.

Matt had brought a gift for him, a hardcover book in German that was published the month before. Matt's excitement over the book made it impossible for him to wait until we were inside to hand it over. As soon as he received it, Berry began leafing through the pages, some of which contained both the German text and glossy pages with works of art.

"Hildegard of Bingen!" His excitement matched Matt's. "One of the greatest intellects in the Western world."

Matt laughed with delight. He had been showing the book to people for a month now, and no one knew who Hildegard was.

"So unfortunate." Berry shook his head. "We have forgotten some of our most important intellectual resources. Without question, Hildegard was a genius of the highest order. The cosmology she expressed in her written works as well as her paintings and music rival Dante in importance. She was the supreme Renaissance human three centuries before the Italian Renaissance."

Matt leapt on this point. Following the lead from the French writer M. D. Chenu, whom he admired greatly, Matt maintained that the true origin of the Renaissance took place as early as the twelfth century and included Hildegard in a major way. She had singlehandedly invented the tradition of natural history, which led, centuries later, to Alexander von Humboldt and Johann Wolfgang von Goethe. "She was so in touch with the divine splendor of nature she invented a new word, *veriditas*, 'greenness,' to capture what she felt about the spiritual riches of nature."

As fascinated as I was by this connection of Hildegard to both Humboldt and Goethe, and even though a dozen questions were burning in my heart, I needed to get these two men out of the cold. I was witnessing their first meeting, and even as it was happening, I knew it was a historic event. Hoping not to extinguish their soaring celebration of Hildegard, I placed a hand as gently as I could on the backs of their shoulders to start them off.

42.

Pierre Teilhard de Chardin vs. Alfred North Whitehead

We had been invited to the apartment of our colleague, Sr. Blanche Marie Gallagher, professor of art history at Mundelein College. When she learned that Thomas Berry would be offering a public lecture, she contacted me and generously offered to facilitate introductions. She and Thomas had been friends for twenty years, ever since discovering their mutual interest in Pierre Teilhard de Chardin, whose cosmology had become a major theme in Blanche's oil paintings.

By the time we reached the Sisters' residence where Blanche lived, night had fallen. The massive brick building was outlined in the dark gray of the day's dying light. For half a century virtually no men had been allowed into the building, and some of those restrictions carried over with the presence of a nun at the door making sure only invited guests passed within. After she found our names on her clipboard, she instructed us to climb the wooden stairs that curved up to the second floor. We would find the apartment at the end of the long hallway. Blanche, a tall woman with flushed cheeks, opened the door.

Thomas and Blanche whooped with joy upon seeing one another. They led the way down the entrance corridor with their arms around each other's waists. At each work of art, they stopped so Thomas could take it in. When he finished, he gestured with his free hand, making circles that grew larger as his hand rose higher.

"So this is where you dream your dreams," he said. "This is where you paint your visions."

As they continued down the hallway, I wrote down his two sentences on a folded-up grocery list from my back pocket. His words surprised me. If I had been the first to enter, I would have said something about the weather or about the difficulty of finding the street address or who knows what. Thomas's sentences transported this ordinary apartment into a sacred cave of dreams and visions. I felt a deep attraction to the moment itself, to the possibility of having more moments like this in my future.

Matt and I took the brown cushioned couch, Thomas the mahogany chair at the end of the coffee table.

"Scotch whiskey for you, Thomas," Blanche said. "Matt, a touch of red wine? Yes? What about you, Brian?"

As Blanche set the drinks on the table, I sat silent, perplexed as to how to start. I might have stared in silence all evening if Blanche had not rescued the situation.

"I have been telling Thomas about your work with Matt," she said. "He would love to hear more. I'm sure I did not represent your thinking very well, but I did my best."

With this as preamble, I blurted out what had been simmering in me for weeks.

"We are doing exactly what you are calling for. Your cosmic story article speaks of our need for a contemporary cosmological myth. That's what Matt and I are creating. We're going beyond the mechanical view of the world Newton invented with its fixed laws. We're throwing out the idea that the universe is composed of dead, inert matter.

"By combining the discoveries of quantum physics with the mystical insights of Meister Eckhart, we present the universe as a magical creative process. That's the main reason we wanted you to come to Chicago, so you could see that we are doing *exactly* what you called for."

"Could be, could be," Berry said. "Tell me, besides Eckhart, what thinkers inspire you?"

"Alfred North Whitehead," I said.

"Very good," he said. "Very good. Whitehead is of monumental importance, perhaps the key philosopher of our time."

His confirmation excited me so much I cut him off.

"Whitehead's conception of matter is so similar to Eckhart's," I said. "Both of them celebrate a world that arises fresh and new in each moment. For both of them, there is no such thing as inert matter sitting there passively. As we have discovered with quantum physics, every bit of matter is vibrating with creativity." I described our work at Mundelein in detail, knowing how happy he'd be to see the fulfillment of his vision. His response surprised me.

"No doubt the work you are doing is of great importance. The interaction between science and theology is one of the more significant events of the twentieth century. Whitehead makes this point in the strongest terms when he claims the future of humanity will be determined by the relationship our generation works out between science and religion. But as important as this is, and as vital as it is for others to join you in your work, I myself am pursuing a different path. I am not that interested in theology or mysticism. What interests me is the universe itself, especially its development through time.

"In my judgment, the greatest discovery of the last four hundred years is the time-developmental nature of our universe. Scientists have come to realize we live not in a cosmos but in a cosmogenesis, a universe developing from a primordial simple state into ever more complex states. It is of such vast importance that contemporary intellectuals don't know how to deal with it. What theologians position their thinking within the context of a time-developmental universe? A tiny number. And even the best of these focus their attention on questions of social justice and largely ignore the natural world. When they do speak of the universe, it's almost always in terms of generalities. Nature for most theologians is only a philosophical concept. They are not speaking of gravity, or natural selection, or the strong nuclear interaction, or any of the other major discoveries scientists have made concerning the universe."

I SAT stunned. He had spoken the bald truth, something I instantly recognized even though I had never heard anyone say it. But instead of embracing this radical statement, I became defensive. I *had* to. His words were sabotaging our work.

"Whitehead is grounded in mathematical physics," I said. "When he speaks of the universe, he is certainly thinking in terms of what modern science has learned."

"No one should discourage your use of Whitehead's thought. As I have indicated, the importance of Whitehead is beyond question. But Whitehead bases his work in philosophy. The vocabulary he uses is not from the particular events of the universe itself but from Western philosophical discourse. He proceeds in this way because he is working out his system by interacting with the central ideas of such philosophers as Locke, Spinoza, Descartes, and Leibniz. For myself, I prefer to work with the empirically verified history of the universe. With the actual phenomena of galaxies, planets, oceans, primates. In particular, I am drawn to the cosmological vision of Teilhard de Chardin, who stays with the language of science as he constructs his interpretations of the universe."

I cut in once again.

"But what does Teilhard have that Whitehead doesn't?" I said. "Their views are the same. Both of them see the universe as creative processes. Both are convinced that everything has a psychic dimension. The only real difference between them is that Whitehead is the better philosopher."

"Teilhard is not a philosopher," he said. "He is primarily a cosmologist, one of the first in the West to contemplate human meaning within a time-developmental universe. As important as it might be to reflect upon the high abstractions of logic, mathematics, and epistemology, Teilhard was more interested in the concrete events of the universe. In pursuit of this, he didn't just study the books of philosophers. He spent decades traveling with teams of scientists to locate and reflect upon the fossils themselves. When he turned to the task of sharing his knowledge, he did not resort to philosophical abstractions, but spoke of rocks and stars and mammalian brains. That is what I mean when I say he stayed with science's revelation of an evolving universe. Teilhard's hunt for the meaning of life always led him back to a developing universe. What fascinated him was the fact that hydrogen became stellar systems with human consciousness.

"There is nothing wrong with staying with the language of philosophy. It is needed. It is of the highest importance. I bring up the issue only to indicate how my own approach differs from Whitehead's. Instead of thinking in the nomenclature of philosophy or theology, I am interested primarily in concrete events, especially the universe's major transformations. These provide the context for my work."

JUST AS when I had read his article on Whidbey Island, my mind was ablaze. Old layers of thought were burned away. Insights from his article returned. Once again, I understood what he meant by "a new cosmological myth." Once again, I was convinced I would never forget it. When Blanche asked if any of us would care for another drink, I asked for whiskey, though I knew the alcohol would be powerless to touch me. I was drunk on what he had expressed. In my euphoria, I blurted out a final question.

"What other thinkers have helped you?"

"The question of whom we should consult in the work of fashioning a new cosmological story is of highest importance," he said. "I hesitate to say too much. This is a period of groping. The metaphor I enjoy most is that of a composer haunted by a music that does not yet exist. Humanity's most significant spiritual challenge is an integral understanding of a universe that is developing through time. We need the sciences, but just as important are the spiritual intuitions coming from the classical civilizations as well as from women's traditions and the indigenous worlds. All of these are absolutely necessary for this work, but we need to keep in mind that by themselves they are insufficient."

Thomas stopped and squeezed his eyes shut, fighting off an impulse to cough. He had done the same several times before, but this time he could not suppress it and the coughing took over. He pulled out a wadded-up handkerchief from his pocket and coughed into this with his eyes closed. The cough came from deep within. I had experienced seeing something as alarming as this only once before, when a friend of Dad's was dying of lung cancer at Lakewood General. The coughing went on and on. Blanche remained calm, watching him. At one point she asked him if there was anything she could do, but he waved her off with his free hand, still holding his eyes shut. When at last it subsided, he kept the handkerchief at the corner of his mouth, speaking now in an even softer voice so as not to bring it back. He picked up on what he was saying as if nothing had happened.

"If we force science's findings into earlier philosophical or spiritual categories, we will unknowingly throw away the best parts. To give birth to a new cosmological story requires a profound creativity. Everything in Western civilization must be put on the table for discussion and reevaluation."

43.

Thomas Berry's Bleak Vision in the Yellow House

For the entirety of his three-day stay at Mundelein College, I followed Thomas Berry around like a hungry dog who knows his master has treats stuffed in both pockets and will be tossing them out, one by one, at appropriate moments. What was needed was constant vigilance so I could gobble up each of them. I escorted him to the student cafeteria for breakfast and lunch, accompanied him to the various classrooms in which he was to speak, took him out to nearby restaurants for dinner with whatever students wanted to join us. Plus—and this was the paradisal climax of each day—our final conversation, just the two of us, at elegant Portinari's restaurant, or rather at the bar, which was separated from the restaurant proper by glass panes decorated with swirling pastel colors.

In the middle of the discussion at Blanche's apartment, the desire to follow him back to New York arose out of some deep place in me. I couldn't bring myself to discuss this as a real possibility. At one point, he asked me about my future plans, and though every cell of my body trembled with a desire to shout, "I want to come to New York and work with you!" I couldn't say it. I couldn't put Denise and Thomas Ian through another move. After dragging them halfway across the continent, the last thing I wanted to suggest was that we go even farther east to New York City. Especially not now. She was exhausted, having given birth to our second child the month before.

THOMAS BERRY's last talk took place on the afternoon of February 14, a Sunday, in a converted mansion right at the place where Michigan Avenue turns sharply to the west, so that when sitting in the house we saw the rushing cars in one direction and the steel-blue water of Lake Michigan in the other. Berry stood before the enormous fireplace and lectured without notes to the four dozen students and faculty crammed into the room with its dark wooden floors and the majestic staircase leading to the upper floors. His main theme was our inability to appreciate the order of magnitude of this historic moment because throughout every field of academia, human thought had restricted itself to domains that were much too small for what was required. Even the most sophisticated thinkers lacked a cosmological orientation, as when they maintained we were entering the "information age."

"That is a grotesque misunderstanding," he said. "Nothing more than clear evidence of our cultural pathology. We have encased ourselves so deeply in the human world we are blind to the biological and geological deterioration taking place throughout the planet. The situation only becomes worse when we realize that the forms of consciousness we evoke in our educational and religious institutions are the very sources of the problems that confront us. Modern societies have fixated on the human agenda, and by so doing they remain oblivious to the planetary dimensions of our disastrous situation."

He spoke with a slight hunch, his face the same face of foreboding I had seen on the conference brochure. When he finished his talk, I turned to facilitate the dialogue with the assembled and saw, to my surprise, Denise standing in the back of the room, holding Baby Sebastian in her arms, Thomas Ian in front of her. I slipped over to the side and made my way back. She whispered she didn't want to interrupt, that I should stay with the group. I said I would return after walking them back to the apartment.

A blast of winter wind hit us face-on when we left the Yellow House. While tucking Baby Sebastian inside my jacket, I asked Denise what she thought of Thomas Berry's talk.

"Bleak," she said. "He makes me wonder why anyone would have children."

"His hope is the new cosmic story," I said. "That it has the power to make things better."

We trudged down Wayne Avenue, facing into the wind.

"You don't need to ask your question," she said. She took my arm and leaned against me as we walked. "Yes. We're going to New York."

"What? No. No way."

"I've wanted this so long," she said. "Starting back in Oregon."

"What are you saying?" I asked.

"That you'd find your teacher. It was depressing listening to him, but he loves Earth so deeply. It fills him. It's even in the sound of his voice. He's the wisdom figure you've been searching for."

Thomas Ian began to cry. He had difficulty walking, which the doctor said would persist until we bought him special orthotics for his shoes. Denise took Baby Sebastian back and I hoisted our older son up to my shoulder to carry him home. The turmoil of indissoluble feelings. Guilt at the thought of moving to New York to study with Thomas Berry contended with adoration of Denise for this immediate expression of extreme love. I knew beyond doubt that this decision would determine the form of the rest of our lives. Its magnitude rose in significance beyond any previous commitment we had made other than that of bringing children into the world. The pathway into the future had suddenly appeared.

44.

The Pilgrimage to Bell Labs

On our two-day drive from Chicago to New York, Denise and I reflected on how we had lived most of our lives on the West Coast. We were born in Seattle, we both chose California for college and Oregon for graduate school. My own lifeline on the West Coast extended north and south into Canada and Latin America in that my summers were spent at Xaxli'p, British Columbia, where my father was born, and my great travel adventure was driving south to Mexico City the summer after high school. But the West Coast was behind us now. We were headed to the center, New York City, the intellectual capital of the world. I had dreamt of going to this strange and powerful land once I learned it was New York that decided who wins and who loses. But I never imagined it would actually happen.

There was one important stop to make. Bell Labs, in Holmdel, New Jersey, was to be my first pilgrimage to the sacred spots of cosmogenesis. The two principal places were Mt. Wilson, where Edwin Hubble gathered the decisive data that demonstrated the expansion of the universe, and Bell Labs, where Arno Penzias and Robert Wilson detected the cosmic microwave background, the light from the beginning of time. I almost missed the opportunity. It was out of our way, an hour to the south, and we were already tired and hungry from day's drive, so when I saw the turnoff to the New Jersey Turnpike, which would take us straight into New York City, I told Denise we should ditch the Bell Labs idea to save time. She argued against that, saying it hardly mattered one way or the other if we took a tiny side trip. I pointed out that everyone was hungry, and Bell labs might even be closed.

"You've been talking about this so long," she said. "Let's just do it."

"Are you sure?"

"Pretend you're from Japan and this is your one and only chance," she said. "You'd feel bad the rest of your life if you passed it up."

WHEN WE arrived at Bell Labs, I was stunned by its size. An enormous, elegant black cube growing out of the New Jersey farmland. Hundreds of offices stacked on top of each other made four solid walls of black looking out at the universe. As we approached and it grew in size, I became ever more dazzled by the thought of actually touching the radio telescope where the cosmic background radiation first entered human awareness. It was in this building that cosmogenesis was established as scientific fact. In the future it would be called the Bethlehem of the new cosmic story. We'd talk about this pilgrimage the rest of our lives, telling our sons over and over again how they had seen the telescope that saw through the depths of time right back to the origin.

Once inside the glass doors, we were met by a woman in a white blouse and blue jacket. She listened to my question with a bright smile. When I had finished, she informed us that public access to the Horn Antenna would begin as soon as renovations were done. For now it was closed off. I told her we were from the West Coast. Could we just peek in? She was sorry, but this would not be possible. Denise was not going to be brushed off. She told the woman I was a scientist, an expert in all this. When Denise finished, the woman turned to me.

"Do you have some identification?" she asked.

I hesitated. The University of Puget Sound had given me a professorship. No point in bringing that up. There was nothing else. I had no proof of any other position because I had no position. In disbelief that her husband was being rejected, Denise tried again, asking the woman if she could speak with the manager. With the same bright smile, she informed us that she was the manager.

I took Denise by the hand and led our little family back outside.

THE MOTEL 6 had a room with two double beds for $29.95. It was just off the four-lane highway, so we could get an early-morning jump on the traffic. While the family settled in, I drove around until I found a Burger King drive through and ordered our dinners. It was embarrassing to be turned away by Bell Labs in front of Denise and Thomas Ian. I had no criticism of the woman. What else was she supposed to do? My truth was out there for everyone to see. I was nowhere. No title. No position. As we pounded out the miles on the freeways over the last two days, I had tried to come up with a name for myself, but none of them were impressive. I liked "cosmic storyteller" but it sounded like something out of a children's book. Every father wants his kids to look up to him. Every father wants to impart to his children the wisdom necessary for success in the world. What was I transmitting?

It was not the first time I wondered about this. While in the midst of our discussions on whether I should stay at Puget Sound, Denise, Thomas, and I had breakfast one Sunday morning at the Hob Nob cafe next to Wright Park in Tacoma. Afterward, strolling toward the car, Denise and I watched in shock as Thomas, two and a half years old, ran full speed into a newspaper dispenser. The impact with the hard plastic knocked him to the ground. He had only just discovered the joy of running but already had decent balance, so it was hard to understand what had happened. Is it possible he did not *see* the box? We ran to him, picked him up. He had a broken nose, but no crying, no tears. Dumbfounded, I asked him what had happened. He said he was running with his eyes closed. When I asked him why, he said he wanted to see what it was like. Nothing more than that. Running at full blast with his eyes closed to see what it felt like.

WE ATE our food sitting on the beds. Thomas Ian, distressed, stayed close to his mother. Moving our residence was yet another disruption for a young child. After setting up in Tacoma where Thomas had friends both in the neighborhood and in his playgroup, we had left that behind and moved to Chicago where, though it took some time, we eventually found a friend for Thomas. Now we were breaking that up and starting over yet again.

To cheer him up, Denise talked about autumn when he would be going to kindergarten and would meet a whole new group of friends. After listening to her, he asked if Baby Sebastian would come with him to school. Not this year, but when he got to be a third grader, his brother would be a kindergartener and they'd be together. Then, as a senior in high school, his brother would be a freshman.

THOMAS IAN burst into tears. It was unlike anything we'd seen. Denise hugged him, asking him what was wrong.

"We won't be a family anymore," he said between sobs. "It's just too sad to say."

Denise took the half-eaten hamburger from his hand and placed it back in the red-and-white paper bag. She folded her arms around him and let him cry, pulling him close to her.

45.

The Great Red Oak in New York City

The three-story Victorian mansion in which Thomas Berry lived had been built just after the Civil War and was on its last legs. It had been condemned and slated for demolition, but the day before the wrecking ball did its work, the Passionist monks purchased it. With intense remod eling they made it into a residence for four of their own, one of whom was Thomas Berry. He was standing on the long porch waiting for me. Denise and I had rented a second-story flat in Mt. Kisco, and this was my first meeting with Thomas since our arrival from Chicago. After a brief handshake he took me around the grounds and introduced me to the community of trees, a sycamore, a copper beech, three ponderosa pines, and his favorite, a gigantic red oak in the back of the house with a commanding view of the Hudson estuary and, farther west, the steep cliffs of the New Jersey Palisades.

After taking in the complex geometry of the oak's thick arms, which spread out in every direction, I asked him how old the tree was. He squinted up at it. Arborists estimated the age between four and five centuries, which would be the middle of its potential life span. He added that Copernicus published his theory that Earth was a planet revolving about the Sun four and a half centuries ago, so the oak tree's life and science's discovery of a heliocentric cosmos began at the same time. Each had begun small and had become monumental. He remarked that whenever he glanced at the tree he was reminded of the entire enterprise of modern science.

We sat in white wicker chairs at a glass-topped table underneath the giant oak. A series of metal rods connected the limbs and supported them from collapsing. It was unnerving. The oak's limbs were far thicker than the black, sticklike supports, and the thought of the whole thing coming down on us hovered in my mind. Maybe by identifying the tree with modern science, Thomas felt confident it was stable enough for us to relax in its shade. I wondered out loud why a tree at midlife would need such rods. His face narrowed. The same arborist who estimated its age determined that it was suffering from a deadly disease with the terrible name "sudden death" and would probably need to be taken down. Thomas had been involved in an ongoing argument about this. He was thoroughly against the idea of chopping up this magnificent being.

HE BEGAN our formal dialogue by asking a question.

"You've made the journey to New York City with its unending activity and inexhaustible creativity. What dream drew you so far from your home in the Pacific Northwest?"

This flustered me. I had assumed he knew. I wanted to shout, "You know why! It's because of your essay! I'm here to learn how to tell the new cosmic story! I thought that was obvious!" It was far from obvious. I had never said this out loud, not even once. It was too much to admit even to myself because Thomas regularly invoked magnificent poets, such as Hesiod, Lucretius, and Dante, whenever he spoke of a new cosmic story. I was not a poet and never would be. Sitting in silence after he asked his question, thinking of what I might say, the truth dawned. All this time I had been assuming he knew about my ambition, when in fact the idea had never occurred to him. I concluded he considered the notion of cosmic storyteller beyond my range of capabilities.

As if he intuited what was on my mind, he spoke directly to my real question.

"The new storytellers will not rise up from science per se. Science will guide the stories each step of the way by grounding us in our best empirical knowledge of the universe, but the foundation for the confidence necessary to become a storyteller is the universe itself. This pertains not just to storytelling but to all roles in society. We find our way into our destinies when we feel we are being commissioned by the whole of things, by life itself. I believe you know this. You've been touched by the universe, even to the point of throwing away your life in order to find your way.

"You will deepen your resolve when you realize that moments such as ours have happened throughout human history," he said. "A tiny acorn gives rise to a huge oak that lives for centuries and then perishes, but not before scattering seeds for the next era. In terms of Western civilization, the transition similar to our own is the thirteenth century when the European Middle Ages were coming to their end. The transformation involved the whole of society and had a bloody dimension as can be seen by recalling the ongoing war between the Christians and the Muslims. We think of Western civilization as being primarily Christian, but after the fall of Rome, control of the Mediterranean was up in the air for centuries. Christian warriors pushed their way into the Near East. Muslim soldiers, after sweeping across Northern Africa, attacked from both the Iberian Peninsula in the west and the Balkan Peninsula in the east. More than once they made their way to the gates of Vienna.

"It is important to remember that these were internecine wars. Both Christianity and Islam came from the same parents, Israel and Greece, and while Christianity brought together the faith of Israel and the mind of Greece in one way, Islam did the same in its own way. They were brothers fighting for control of the kingdom.

"In the midst of this upheaval, the pope called Thomas Aquinas to Rome to deal with this threat at the theoretical level. Christian thinkers were at a disadvantage in confronting Islamic scholars in the sense that Europe had only recently discovered the thought of Aristotle. Aquinas's first act was to order fresh translations of all the Aristotelian texts. Drawing upon these and traditional Christian scholarship, he reinvented Christianity by incorporating Aristotle's comprehensive science and metaphysics. His *Summa Theologiae* provided medieval Europeans with a cosmic story powerful enough to hold the community together. It provided answers to life's recurrent questions. They knew who they were. They knew why they existed. They knew what was good and what was evil.

"These orientations of Aquinas, at one time the sinews of Western civilization, lost their power when modern science entered history. This happened not just in Europe but throughout the planet, in every major civilization. The brilliant visions of Asia, Africa, and India were reduced to the category of mythological stories. For many, the two world wars were testament to the truth of Neitzche's observation that the Christian European God was dead.

"Now we come to our time. Both the challenge and the threats of our moment are orders of magnitude greater than what Aquinas was dealing with. The situation is ambivalent in the extreme. Science gave birth to our technological power, which we are using to rob Earth of its vitality; but science has also discovered that the universe as a whole is developing, which as you know I regard as the primary revelation of our time, at the scale of a world religion. At risk now is not just Western civilization but the wildness and beauty of Earth itself. Our greatest hope is to meet this challenge by telling an integral, cosmological story, one that will guide us into a future flourishing with life."

Too excited to hold it in any longer, I shouted, "I love this! It's what we have to do! It's what I *want* to do!"

He smiled, then looked away again but not before I saw a glow in his eyes, as if he'd been waiting for that.

"This is the northwestern corner of the Bronx, not the Vatican," he said, his voice even softer. "I'm a historian of world history, not the pope. But perhaps I wrote my article for the Whidbey Island community so you would come to New York."

He unzipped the black leather satchel lying on the tabletop and pulled out Teilhard's major work with a blue cover battered from use. He slid it toward me.

"Please ignore my comments in the margins. I have an unread copy somewhere, but I haven't been able to put my hands on it. The great achievement of Teilhard is to recognize that science has discovered cosmogenesis. The developing universe provides the comprehensive context for understanding in that to know something has come to mean to know its history, to know its development through time. There is no fixed and final form of Christianity, there is instead the history of Christianity. This is true for science as well, even though we tend to identify current scientific understanding with science in general. But science is also evolving. Everything is; even the universe as a whole is evolving from its relatively simple beginning to more complex forms.

"The universe is coming to know itself. That is the meaning of the human species. The meaning of our existence is to provide a space in which the universe can reflect upon and activate itself in conscious self-awareness. This can only be achieved by a unified humanity, which Teilhard called the noosphere. The early Earth in the form of molten rock gave rise to the atmosphere and oceans. Over another billion years of evolution, the atmosphere with the oceans, minerals, and sunlight gave rise to the biosphere with all its fantastic diversity of life-forms. And now, out of that complex network of relationships, the Earth, through *Homo sapiens*, has come to know itself. That is why we are here.

"It sounds strange to modern sensibilities to define our humanity in terms of the universe, but such cosmological orientations were common in premodern cultures as well as in the work of Schelling and Bergson. Telling the story of our time-developmental universe so that the universe can enter into awareness of its complexity and beauty is the greatest spiritual task of our time.

"Which is why you struggled to fit in at the University of Puget Sound. It seems to me your role is not in science exactly but in cosmology. Your destiny is to tell the universe story."

As I drove back to Mt. Kisco, I had to keep reminding myself that the speed limit was fifty-five. I was racing to tell Denise. At my very first meeting in New York City, I learned who I was and what I was about.

46.

Terror in a Vast Universe

As a way of surrounding myself with the story of the universe, I tacked a poster of galaxies to the wall of our living room. This computer-generated image was based on observational data of the nearest one million galaxies as photographed by astronomers at Lick Observatory in California. It was the first such visual depiction of the large-scale structure of the universe, designed to give a feel for the three-dimensional shapes of the galaxy clusters. Built upon hundreds of thousands of hours of painstakingly detailed work, it was available for $12.95 from the Astronomical Society of the Pacific.

Its presence in our upstairs flat made a difference right away by providing my first opportunity to tell the story of the universe to the general public, in this case just one person, Margaret Vodanovich, the friendly lawyer from across the street. As Denise brewed tea in the kitchen, Margaret and I were left alone in the living room. Pointing to the Lick Observatory poster, she asked what she was looking at. I got out of my chair to explain.

"Even though the picture looks like a random scattering of white dots,"

I said, "these are exact locations—based on telescope observations—of the nearest one million galaxies. In order to get a lot of data into the picture, the astronomers squeezed twelve galaxies into each white dot, so there's really only eighty thousand dots to represent the million galaxies." I glanced back at her. Her face had a look I could not interpret. I hated boring anyone. Maybe I needed more enthusiasm.

"Just think of it!" I said, pointing at the map. "Each little dot represents a trillion stars! Can you believe it? Each *dot*! In terms of volume, a dot is a trillion cubic light years. Which means that even traveling at the speed of light, it would take more than a hundred thousand years to cross a *single dot*." As I glanced back at her, the situation with her face had deteriorated. It wasn't boredom. It was more like disgust. Whatever it was, it drained off my excitement. "The real kicker is that these million galaxies represent less than a tenth of one percent of the known universe. Not that we know how many galaxies there are, but even so . . ."

I stopped speaking. Margaret had stood up and was walking away from me toward the staircase that led downstairs and outside. Denise appeared in the kitchen doorway holding a tray of cups and a teapot and gave me a "what's happening?" look. I followed Margaret with my stomach churning. I had no idea what I had done. I called her name but she kept going. I followed her down the stairs.

"Margaret, please, what is it?"

"I can't handle that," she said. She was working the latch on the picket fence. She spoke over her shoulder. "Why are you fixated on destroying my faith?"

She crossed the street and disappeared into her house, the white screen door creaking slowly to a close behind her. I was certain she would love learning all this. What had just happened? I felt the queasiness that comes right before throwing up.

IT WAS the first thing I asked Thomas when we met on Saturday. We drove to the Broadway Diner, in the northwest corner of the Bronx, on 261st Street.

The management had installed a juke box at the far end of the counter featuring songs from the early years of rock and roll. Our two menus sat closed up on the Formica-topped table of our booth. I was following his lead, ordering a cup of coffee and supposing we would get our lunch later. As soon as we were settled in our booth, I recounted Margaret's desperate exit. His response was immediate.

"It's not always possible to reach your listeners."

"What would you have done?" I asked.

"I would have told her the story."

"But I was!"

Thomas looked at me out of the corner of his eye.

"Are you sure?" he said with a smile. "In a sense, of course, you were telling the universe story in that you were situating your listener in the great array of galaxies, but in another sense you weren't telling a story. You were analyzing a snapshot. An examination of one frozen image of the story only supports the assumption that the universe is a conglomeration of objects. Think of that poster of white dots as akin to a hundred butterflies pinned to a corkboard. What one wants is not a collection of insect corpses, but a group of living butterflies flitting about a forest. That is the power of story. Only story can give a sense of what could be called the dance, or the dynamism, or the spirit of things.

"We have detected the creativity of cosmological time. That is the key revelation. Primordial plasma broke into a trillion galaxies and complexified further into the splendor of the Earth Community. That's the drama we've been presented with, and it was the last thing we expected to find. The classical civilizations regarded the cosmos as a whole as unchanging, but we today find ourselves in a developing cosmogenesis. This is a thrilling revelation with enormous promise for unifying humanity, but we cannot pretend that this story is small or tame in its dimensions. It's terrifying. It will not be domesticated. There is a grostesque aspect pervading cosmic evolution that cannot be covered up or ignored. Nor should it be. The opacity of the story is essential for the work of deconstructing our false assumptions.

"You yourself have already begun this process," he said. He spoke with such a soft voice I had to lean over the table to catch the words. "In your study of the universe, you took the first steps as the stars awakened a glimpse of your cosmological being. You then attempted to fit this larger sense of yourself back into industrial society and found the impossibility of that. That's when your career as an academic scientist came to an end. Such experiences will become more widespread in the next century. Your own alienation involved your realization that the modern university is out of alignment with Earth's dynamics; others will learn the same thing, but via our legal systems, or our forms of governance, or our economics, or our religions. The suffering to be endured is immense. We will see a steep rise in families breaking apart, in alcoholism and drug addiction, and though we will try to understand these on a personal level, the ultimate cause will include the inadequate forms of our fundamental institutions. They are too small to deal with the cosmological nature of what is taking place.

"In your role as a teacher you will be tempted to think we can carry this through in our lifetimes, but that will not happen. The challenge is far greater than what can be accomplished by a single generation. You will also be tempted to divide the world into those who see the truth and those who do not. That too must be fought against. Always bear in mind that not everyone has the freedom to refuse to participate with forces destroying the planet. And not everyone has friends who are on this journey with them. Remember these as well as other ways in which you have been showered with grace. That is why you yourself have a chance at succeeding," he said.

"Succeeding at what, exactly?"

He nodded.

"Awakening a new era. The exciting action of our time is the creation of a cosmological form of humanity."

"If I hope to become a cosmic storyteller, what is my next step?"

"Come to understand you are not a scientist. You are not an American. You are not male. You are not even primarily a human. First and foremost you are a cosmological being."

"Which means I'm . . . ?" I hesitated.

"The universe. In the mode of a human."

47.

The Supreme Poet of the Medieval Cosmos

Thomas proposed Dante as the next step. His point was that if I were serious about becoming a cosmic storyteller, I would need to underline the wisdom dimension of science's narrative of time, and the best way to begin this task was with Western civilization's most celebrated spiritual cosmology, Dante's epic poem *La commedia.* His choice of Dante was due to my birthplace. The movement of humanity into a new cosmology will be rooted in unique forms depending on one's place. If I had been born in Onondaga Nation or in North China and I wanted to tell the cosmic story, I would focus on the stars and galaxies and the expanding universe, but my interpretations would come from the distilled wisdom as lived and reflected upon in those cultures. As I was born in the West, I needed to find a way toward a new cosmology using the thought traditions of the West.

Thus it was that on November 13, 1982, Thomas began his instruction of how I could benefit from a study of Dante. He was dressed, as always, in his brown corduroy sports jacket. We were in the library, alone, surrounded by hundreds of volumes covering the classic works of Chinese and Indian and Western philosophy. The morning light streamed in through the two windows. Birdsong filled the room, which Thomas located in a brilliant red cardinal on a lower branch of the maple tree just outside the solarium. Thomas began by asking me a question.

"What is the year of your birth?"

"Nineteen fifty," I said.

"Yes, that's what I remembered this morning. It came to me that in terms of chronological time, the relationship of Dante to Aquinas is approximately the same as the relationship of your generation to Teilhard. There's a brief overlap in both cases. Dante was born in 1265, which was when Aquinas was embroiled in theological disputes in Paris and Italy. Though Dante never met Aquinas, there is speculation that Dante studied Aquinas's philosophy with one of his students, Remigio dei Girolami. Girolami had worked with Aquinas at the University of Paris before becoming a lector at Santa Maria Novella in Florence, where Dante lived.

"As we have discussed already, our moment in history has a strong parallel with their situation in the late thirteenth century. St. Thomas's *Summa Theologiae* provided a new cosmology synthesizing Greek science with Christian faith. If not for Dante, this vision of Aquinas would have remained shut up in the scholastic Latin of the monasteries. Our situation is similar. Charles Sanders Peirce, Alfred North Whitehead, and Teilhard have constructed cosmological visions that present human and cosmic meaning within a developing universe. Unless we want to leave their work on a library shelf in the university, what we require is the cosmic narrative in the forms of film, music, art, and ritual. It needs to be told as an epic and brought into world literature. That's how it'll come alive on every continent and in every culture. These differentiated expressions of the universe's creativity will empower our species to reinvent itself."

As a way of introducing me to Dante's poem, Thomas began by telling me that I should think of *The Divine Comedy* not only as one of the world's most elegant works of poetry, but as Dante's personal narrative of his journey from misery to bliss. At the start, Dante is confused, but as he learns the ways of humanity in his visit to hell, and as he undergoes the transformation of consciousness in purgatory, he soars into the celestial realms to witness the exalted states for which human beings are destined.

I was thrilled by the whole idea of studying Dante in order to tell the new cosmic story. As sophomores in high school we had read an abridged version of the poem, and even though I had forgotten most of it, I did remember how impressed I was by the thought that Satan lived at the center of Earth. This was English class, and it matched the period we were studying in history, the Middle Ages, which was fascinating for us students, kings and queens, palaces, medieval battles, and the belief in heaven and hell. I was astonished to think medieval people believed a spiritual being lived underneath their feet. As a way of sharing my excitement, I told Thomas I had been amazed by the thought that people like Dante thought Satan was frozen in the center of Earth. Thomas said nothing. I asked him directly whether people in the Middle Ages believed that.

After another hesitation, Thomas spoke in his soft voice.

"Dante didn't think Satan was at the center of Earth."

"Really?"

He nodded without looking at me. My question had embarrassed him.

"Oh," I said.

I could feel my cheeks grow warm as I realized I was out of my depth. Strata of understanding were at play of which I knew little or nothing. I resolved to listen as carefully as I could and to say as little as possible. As if Thomas sensed my withdrawal, he changed directions.

"There's a faint resemblance between Dante's approach and your own."

"Really?" I said again.

"But before I get to that, we need to identify Dante's relevance for our time. For centuries his work served an entire civilization with its magnificent depictions of good and evil, of holiness, of how we miss the mark, of what it means to reach one's deepest fulfillment. Over the last century, Dante has been pushed to the margins. The cosmology of his poem is no longer a viable vision of the universe. We need to know why."

48.

The Universe Takes Flight as a Red-Tailed Hawk

After a month devoted to a careful reading of Dante's poem, Thomas laid out his basic idea.

"The simplistic criticism of *The Divine Comedy* is that Dante presents his vision in a geocentric cosmology in which all the stars and planets and even the Sun revolve around Earth, a vision overturned by Copernicus's theory in 1543 that placed the Sun at the center of the solar system. All of this is true, but there is a deeper inadequacy that has to do with time. For Dante, as for most classical civilizations throughout the planet, the creation of the universe took place at one time in the past. As Dante makes his way down into the inferno and up into the realms of the planets, the only thing in the universe that changes in any significant way is the human being. Everything else is fixed. Everyone else fits into an order that has existed since the beginning.

"But modern science has discovered that everything in the universe is in movement, everything is evolving. The stars. The galaxies. The planets.

That's why we say that medieval Christendom, as well as the other classical civilizations, lived in a *cosmos*. We don't live in a cosmos. We live in a cosmogenesis, a universe that is becoming, a universe that establishes its order in each era and then transcends that order to establish a new order. This is why Dante's work, which is unsurpassable in the realm of poetry, misses the essential challenge of our time. Dante's focus is the human individual and his journey to salvation. As important as that might be, it is only a part of what we are dealing with.

"In the new cosmic story, the hero is the Earth Community itself. And the hero is the individual human. It's both. From one perspective, the Earth Community is the outer form of the individual human person; from the other direction, the human person is the Earth Community reflecting upon itself in conscious self-awareness. We cannot speak of the spiritual development of a person while the planet as a whole is withering away. Either the person and the community blossom forth into their full destinies together, or they both collapse into a degraded state beyond the worst horrors Dante was able to imagine.

"Which brings us to the point of similarity between Dante's vision of things and your own. Both you and Dante consider the happenings on Earth to be of supreme value. Dante's way of expressing this is geometrical. He conceives the universe as revolving around Earth. It is on Earth that the most significant activity takes place, which for Dante is the spiritual development of humans. That humanity is of central importance in the universe is understood by Dante's angels. Even one act of love on the part of a single human results in a chorus of praise by the entire community of saints and angels in Dante's heaven.

"In our conversations, I have come to see how you too accord Earth a supreme value, which you express not via geometry but creativity. Throughout fourteen billion years, the ultimate power in the universe has constructed so many things, but nothing more complex than the Earth Community. As you say, there may be other living planets, but in the *known* universe, in the universe about which we have empirical evidence, Earth is more alive, more intelligent, more beautiful, more sensitive, more complex than any galaxy or star or planet we have examined. Earth then can be considered the primary revelation of what the universe is aiming to accomplish.

"I would say that your difference with Dante is that you recognize every particular as having dual dimensions, the individual level and the universe level. A red-tailed hawk, at its local, personal level, is making decisions moment by moment; but the action of its flight reveals the universe as a whole in the curvature of space-time that influences its motion through the air, in the supernovas that constructed the elements of its body, in the million experiments of its ancestors that have shaped its feathers. From the perspective of the individual, we see a particular hawk flying above us. From the perspective of the time-developmental universe, we behold fourteen billion years of creativity taking flight."

I stared at him. Gawking would be a better description. He had devoted a lifetime to the study of the world's wisdom traditions that surrounded us here in the library, but his search had spilled out in every direction. Upstairs were a thousand volumes exploring the wisdom of women's and indigenous traditions. On the bookshelves of the solarium were his geology texts. What was once a second pantry now housed his biology and scientific cosmology volumes. On the third floor were the shelves housing history texts. It was his lifelong reflection on the history of philosophy and religion that had led him to his insight concerning the significance of a new cosmic story arising from four hundred years of scientific study of matter. He realized that something foundational was taking place as this new story took hold of the human imagination.

"Let me be honest," I said. "As much as I would like to claim you have described how I think about red-tailed hawks, the truth is, I doubt I've ever thought like that."

"You do and you don't. It's there implicitly in your thinking, just below your conscious mind. But like the rest of modern society, you struggle to keep the universe dimension in mind. In our drive to analyze entities into their component parts, we push that aside. That is why our default mental state is to see a universe filled with objects that we have catalogued by the millions, an achievement we must not underestimate. But this analytical work comes with a cost. This deformation of consciousness is not identified in any psychiatric manuals because the writers of such manuals are themselves victims of it. When the time comes for us to include it in our medical nomenclature, it might be called the 'anti-cosmological pathology.'

"To varying degrees, this psychic condition characterizes most individuals throughout our advanced civilizations." He smiled. "And this includes those who wear corduroy jackets and give lectures on it."

49.

Plagues of Symbolic Consciousness

Thomas lifted his finger as if remembering something. Chuckling to himself, he pulled out three small books from his clothing, one from inside his brown corduroy sports coat, the other two from his black slacks, and placed them on the table. These three tiny volumes were his personal copy of Dante's poem, which he carried with him wherever he went. They looked like moleskins that had been purchased half a century ago. The worst of the three had additional wear. Either Thomas had read that volume the most, or it had suffered a cycle or two in a washing machine.

"Is that Inferno?" I asked, pointing to the beat-up pocketbook.

"Paradiso," he said. "I read Dante backward." He scratched the hair behind his ear with a nervous twitch of his fingers. "I needed courage. I was nearly catatonic over the state of things, both at the human and planetary levels. The spiritual journey is difficult because the first step is recognizing the ways one's character is out of sync with reality. For Dante, to be out of sync is the pathological state of desiring that which degrades us. That was my fear. That on some basic level I was creating misery in the world without being aware of it. I stayed with Paradiso until I felt I had built up the necessary strength for the spiritual journey.

"The most terrifying moment in Inferno is Dante's encounter with Medusa, a Gorgon whose hair is composed of venomous snakes. In classical literature, one glance from Medusa was enough to turn one into stone. This figure is Dante's way of showing us the vulnerability of the human mind. Dante uses Medusa to convey the terrible truth that when we reify false theories, we turn our minds into stone. Frozen thus in a pathological vision, we confuse evil with good. In that degraded condition, a human has scant hope of finding a way to align with the cosmological order of things.

"Dante's insight is that certainty is a dangerous state. Our view of the world might be correct, but there is no method that can determine this. That sort of proof is not given to humanity. We have to work out our understanding of things always in the humble awareness that final knowledge is not an option for us.

"On the other hand, we *can* determine when our behavior is out of alignment. That's the human condition. We can't know when we are right, but we can determine when we are wrong."

"How is that?" I asked.

"Charles Sanders Peirce provides the test. If our actions bring about the opposite of what we intend, there is a very good chance we are the unfortunate victims of a reified falsehood."

"What would be an example?"

"Industrial society sees the universe as a collection of objects. That view has frozen modern consciousness. But we find ourselves in the midst of the worst destruction of Earth's life in sixty-five million years. We have brought about a mass extinction of life. No sane person desires this. No sane person wakes up with the intention of destroying all life. And yet here we are, doing just that. If Peirce were here, he would point out that the results of our actions are the opposite of our intentions.

"Because of our habitual forgetting of the universe as a whole, we do not have the intellectual capacity necessary to understand what it means to extinguish a species. The major philosophers in the classical civilizations were equally incapable. Plato, Confucius, and Shankara all understood that an organism could die, but none of them imagined that birth itself could die. Extinction did not fit their cosmologies. Even thinkers of the magnitude of Ralph Waldo Emerson argued extinction was an impossibility because nature, so he believed, was too robust to be destroyed at the species level. But while the philosophers argued over whether or not a single species of life had ever gone extinct, our industrial way of life was destroying thousands of species each year. Perhaps millions will disappear because of us.

"Even though Aquinas knew nothing of the organic evolution of life, his theology takes a first step into understanding the magnitude of what is taking place. Aquinas asks a simple question: Why did the divine give birth to so many forms of life? Why not one form of life? After all, birds give birth to baby birds. Elephants give birth to baby elephants. His answer is that creation in all of its diversity is necessary to provide a full revelation of divine reality. No one form of life by itself is enough to convey the richness. In Aquinas's theology, then, to eliminate a species is to eliminate divine presence. To devastate a beauty that required billions of years to emerge is an act of what we can call biocide. For Aquinas, the more accurate term is 'deicide.' We are condemning all future humans to live in a universe that has lost its sheen, a universe less capable of awakening the humane or spiritual qualities of its children."

THOMAS HAD finished his answer to my question. I sat in silence, stunned by the revelation that we were causing the first mass extinction in millions of years. How did I not know this? What other gigantic truths was I ignorant of? I felt horribly confused. What could be more ludicrous than someone telling the story of the universe without mentioning the unraveling of Earth's life?

We sat in uneasy silence. I tried to come up with something to dispel the pall. I could think of nothing. I felt only deadness. When I was in grad school, Oregon farmers, as a way of replenishing the soil, burned the stubble in their fields, filling the skies for days with billowing black clouds. When weather conditions were unfavorable, as when the winds were blowing in the wrong direction, soot settled down as a light gray ash. All the colors of the houses, trees, and lawns were replaced with uniform gray soot. That was me, sitting there as this horrible news descended on me.

AN HOUR later, I was at Howard Johnson's, waiting for Lennie Auclair, a student from the year before at Matt Fox's institute. I wanted only to go home and tell Denise what I had learned and talk it out with her. But there was no way of contacting Lennie while he was driving from Chicago through New York and on to Maine. He strode in and glanced about, rakish good looks, five o'clock shadow. When he spotted me he bowed with mock grandeur. An actor and playwright, he had studied at Ringling Brothers to become a professional clown, following in the footsteps of Ken Feit. In order to work with Feit, Lennie had applied to Mundelein College where Feit was to teach. Death had destroyed that dream.

Though the terrible facts from Thomas Berry suffused me entirely, I could not begin to express any of it for Lennie. It was too vast. I could not get distance from it enough to speak of it, especially with the strange thought that in my ignorance I was part of the cause. Would Lennie benefit if I told him? Would anyone?

Halfway into our meal, Lennie remembered a favorite Ken Feit playlet. With a pair of scissors and a notable dexterity, Feit would make cuts into a tightly folded piece of paper, then unfold it to reveal a delicate little horse with a horn between its eyes. He made this creation dance about for a moment or two then plunged it into the candle's flame where it caught fire. Feit's happy face transformed into a mask of horror as the delicate animal burned to a black crisp. After cradling this in the palms of his hands, Feit, with a huge exhalation of breath, blasted the remains into tiny fragments.

"Did you ever hear what Robert Bly thought of this?" Lennie asked.

"They were doing a gig together in Chicago. Right after the unicorn bit, Bly yelled out, 'Feit, you're in love with death.'" Lennie shrugged his shoulders. "Maybe he was."

Death was coming at me from every direction. I hated hearing that Bly had publicly criticized Feit like this. So what if Feit was in love with death? What did that matter if life-forms all around the planet were disappearing? And it was happening now. I had nothing to say about it. I was in the dark. Lennie told more of his stories, which were usually funny and sometimes profound, but I couldn't hear a thing. My eyes had glommed on to the bubbles rising in my glass of beer. Like the early cloud chambers used to detect elementary particles invading Earth's atmosphere, the bubbles traced out lines. Cosmic rays. Coming from distant supernova explosions. When a star blasts itself to smithereens, the violence makes its long journey to us, right here, messing with this glass of beer that had hitherto imagined itself safe from any such intrusion. The planet was already dying, and it could get much worse, even one second from now. If a supernova exploded within two hundred trillion miles of Earth, even if the explosion had happened years ago, its intense gamma rays could be one minute away from reaching us. Though they roared toward us for years, we would know nothing of their existence. How could we? Even if we accidentally trained our telescopes in their exact direction, we would see nothing but a star peacefully twinkling. The blast of light from the explosion would not have reached us yet. Then it would arrive. A wave of incineration would instantly destroy the ozone and kill off the phytoplankton. All the oxygen would be vacuumed from the atmosphere. A trillion carcasses would stack up on an Earth reduced, in a geological instant, to a husk. I sat with Lennie and knew this cosmic slaughter could commence in the next minute.

If it didn't come in that minute, it could come in the minute after that. There was no way out of this fearful state.

50.

The Magnanimity of Sunshine

Thomas served as president for the American Teilhard Association, which held meetings once a month. The schedule of events started in the early afternoon with a presentation on the new cosmology followed by discussions among those assembled until early evening, at which point a light dinner and drinks were offered to the group. I arrived early to help Thomas with the food shopping. Inside Safeway, I pushed the cart while Thomas marched up and down the aisles, plucking out the items that would be served for dinner. Heads of lettuce, russet potatoes, three whole chickens, and four cans of Campbell's mushroom soup, which he told me with a laugh was his secret ingredient for almost every meal he prepared.

Even though we were in a bit of a rush, we had to make one more stop, a special bakery Thomas loved for its apple pies, so instead of heading back to Riverdale, we drove south on Palisade Avenue toward Manhattan. I decided to bring up a question that had been bothering me. I had avoided asking it because I knew it would reveal another hole in my understanding. But after my embarrassment over the galaxy poster, things could hardly get worse.

"You speak of matter as having a spiritual dimension," I said. "But you also say that the universe story is the discovery of science. I don't see how these two can go together. If we are using science to get at the truth of things, we can't use a phrase like *spiritual matter.* How do you reconcile this?"

"What you say is accurate. Though rooted in the empirical details of contemporary science, the new cosmology is not science. It's an interpretation of the data science has given us."

We slowed to a stop at the red light. Even as we waited, he kept his eyes on the road ahead.

"The key assumption of the foundational scientists of the West, including Galileo Galilei and Isaac Newton, is that the universe is out there and that the human mind can examine it and discover the laws that determine its actions. Our whole approach is based on this belief of our separation from the natural world. The belief that we are inside our brains, and from that perspective we can understand the things out there. This dualism was given its philosophical foundation by Rene Descartes and its operational method by Francis Bacon. It has led to spectacular knowledge of the universe. But it has come to its end. The irony is that its success has shown the falsehood of its enabling assumption. When scientists discovered cosmic, biological, and cultural evolution, they demolished the notion that we are ontologically separate. We are not separate from the universe. The universe and Earth *constructed* us."

"This is what you mean when you say we humans have to stop hogging all the noble qualities for ourselves. Things like love and generosity and spirituality."

Thomas reached over and tapped my forearm with his index finger. He was smiling.

"Simply that."

"So you would be happy saying something like, 'The Sun is generous'?" I asked.

"Certainly."

"But still, how is that not anthropomorphism? That's what I would be accused of."

"The real difficulty is the diminished form of modern languages. In English, 'generous' has come to be identified with a particular action of a human being. We've imprisoned the word. With that stunted meaning, it is of course nonsensical to say the Sun is generous. What we need is to expand English to make it capable of giving expression to our deep experiences. The Sun is generous, but not as a human being is generous. The Sun is generous as every star is generous. It transforms its mass into light. What are the numbers again?" he asked.

"Four million tons."

"Four million tons of the Sun are converted to light?"

"Yes."

"And that takes place each second?"

"Yes," I said. "Four million tons of the Sun become light in each second."

"Astounding," he said. "With that in mind, we can ask ourselves a new question. If in each second, the Sun is deluging us with light, and if that light has powered every act of human love and generosity since the beginning of humanity, shouldn't we be free enough to call this bestowal of light 'generous'?"

I FELT giddy, alive. A weight lifted from me. No, it was more than a weight. A metaphysical torture instrument had come undone, something I now realized I had endured for years. The spikes had come from so many directions, including Dyson's proclamation, "It's just a big ball of gas." It was a feeling that brought back a similar feeling when Matt Fox spoke of Hildegard's art as revelatory. Both Matt's and Thomas's words opened up a passageway. Perhaps this was what Dolores was saying concerning Plato's cave. That it was possible to escape.

All I wanted was to stay here in the light that washes over anyone who steps out of the cave. To set up a tent, like Peter the apostle. My fear that I would be pulled back inside the cave forced me to ask a deeper question.

"But there is a difference," I said. "It's a question of *intention*. Humans *intend* to be generous. The Sun doesn't have any choice in the matter, right?"

WE HAD arrived. Thomas parallel parked the car across the street from the one-story brick building. Faded red letters on the awning did their best to identify the bakery as KEISHA'S SWEET AFFAIR. When we got to the door I noticed I'd left the back window down. I asked Thomas for the car keys. He squinted.

"The groceries," I said.

He broke into a smile, tossed his hand at me.

"If they need them that badly, they *should* take them."

51.

The First Basic Law of the Universe

The tiny store had two banked glass display cases with trays of glazed pastries. The baker, rail thin, severe face, her bald head signaling ongoing chemo treatments, worked the cash register. As she handed over the white paper bag to the only other customer in the place, she turned to us. Her face lit up.

"Father Thomas!"

"Keisha!"

She edged between the two cases and threw her arms around him. When she broke off the hug, she saw she'd imprinted Thomas with flour. Even though Thomas laughed and told her not to bother, she employed a white cooking cloth to brush him off, ignoring his protests as she would a child's. When Thomas said I was from Washington State, her eyes grew wide. She claimed she was the world's number-one fan of the SuperSonics, Seattle's basketball team. Before I could relate my tiny connection to the SuperSonics, her eyes went wide again and she disappeared into the back.

She returned with a sheet of paper held above her head, predicting that Thomas would be surprised. We were instructed to sit down, as the pies were not quite done. Thomas handed me the page when he was done reading. As I looked at the string of A's on the first-semester grade report of one Cornell Williams at Howard University, Thomas gave the backstory.

He had met Keisha at the funeral Mass for her husband several years back. Thomas was presiding at the Mass only because his colleague, Neal Sharkey, had fallen sick at the last moment and couldn't find a replacement. Thomas himself rarely, if ever, performed the clerical services of the priesthood. He explained that he had entered the monastery in order to brood on the nature of the world, and the only two places he knew where this could happen without interruptions were the monastery and the penitentiary. As a monk, he could devote himself full time to study. His decision to become a priest had less to do with performing religious rites than with the opportunity to ponder the meaning of existence.

Though he was unpracticed at giving sermons at funeral Masses, Thomas figured he must have said something that mattered because several weeks later Keisha came to him with a request. Would he speak to her son? Would he bring Cornell back to his happy self? So in addition to being the chair of the History of Religions program at Fordham University, Thomas was now a therapist for a teenager. Cornell visited every Saturday, pedaling over to the Riverdale Center on his bike. Having no professional training in counseling, Thomas proceeded with Cornell as if he were a graduate student, inviting him to his various discussion groups on history or cosmology or spirituality. In a matter of weeks, by dint of natural ability and hard work, Cornell blended in. Thomas's conclusion was that the fifteen-year-old mind was capable of understanding anything.

This is where Carl Sagan entered the story. As was his habit, Cornell expressed his frustration with being stuck working at a bakery. He wanted to get away, to go where the "really important things" were happening. Thomas told him that without question he would soon be striding forth into the world, but if Cornell couldn't see the important things taking place in the bakery, he would struggle seeing them anywhere. An idea came to Thomas.

He recommended Cornell watch the *Cosmos* series, that he should try to understand what Sagan meant by proclaiming, "If you want to create an apple pie from scratch, you must first invent the universe."

Cornell did watch the series. Something clicked. He started to see his world in a new way.

On multiple occasions, Keisha had offered to pay Thomas for working with Cornell, but Thomas would have none of it. She eventually came up with the idea of supplying apple pies for his symposia. And thus the tradition began, and as much as I appreciated it, I was also dismayed by the story's culmination in Carl Sagan's television series. I couldn't have a conversation about the universe without his name coming up. Once again, all roads led to Sagan.

THE *COSMOS* series had premiered two years before, in my final year at the University of Puget Sound. I was simultaneously attracted and repelled by the shows and gave up watching after three or four episodes. But turning off the television was not enough to block his presence. Students in my classes laughed if I said the word *billions*. They compared the way I dressed with Sagan; whether similar or dissimilar, it didn't matter. In our seminar in mathematical cosmology, small criticisms of his science cropped up. Even Sheldon got into it, proclaiming it was inevitable Sagan would use the entire cosmos as his backdrop. As he put it, "I too learned by age five I was the center of the universe."

All of this left me with an awful, unsettled feeling. I was glad when the broadcasts were finally over, but even then, the collisions with Sagan continued. In the spring semester after I had decided to leave Puget Sound, I was contacted by Carrie Washburn, special assistant to the dean, who asked if I would teach a summer course based on *Cosmos*. She was thrilled that the university had obtained the rights to use the series as part of a course. She obviously didn't know I was resigning from the university, and I didn't want to bring it up with her, so I said I was too busy. She resorted to flattery, telling me it would be difficult to find someone who would, as she put it, "not be intimidated by Sagan's ego." I was forced to tell her the truth. That I was falling apart. That I had resigned from the university. That my future was blank. She listened and said, "I know." Which showed the depth of her devotion. Her desire to share Sagan's work with the students was so great it steamrolled any concern she might have had about my struggles.

WHEN THOMAS finished his story, I followed with an account of my uncomfortable relationship with Sagan.

"It's just envy on my part," I said in summary.

"Do you know Sagan personally?" Thomas asked.

"We had one letter exchange."

"You speak of envy. Do you resent him?"

"Not really," I said. "He taught science to the world, which is so great. But it's like there's nothing left to do since he's already done it. It might sound like sour grapes. Maybe it is sour grapes. But at the same time, there are some things not right . . ."

"You've heard me speak of the first basic law of the universe?" he asked.

"Differentiation," I said.

"Yes. From the very first instant of the universe's flaring forth, matter seeks to differentiate. Very quickly, plasma transforms into a trillion galaxies, each one unique. It's the most amazing feature of existence. Think of what it means to say that no two galaxies are the same. This ability to generate novel creations is considered divine when it appears in humans as in Shakespeare's invention of all his characters. Or Murasaki Shikibu. Her *Tale of Genji* with its hundreds upon hundreds of personalities expresses this law of differentiation in a superb way. As time rolls on, it never ends. The trillion unique galaxies give rise to trillions upon trillions of unique stellar systems. Each era of the universe is unique, never to be repeated. It's as if at dawn a deep and resonant voice calls forth to every entity in existence, 'Become your unique self. The universe advances only if you blossom forth as you.'

"If differentiation is the first law of the universe, *sameness* is a cosmic crime. Sameness is the one thing that will not be tolerated. At birth, each of us is promised a quantum of energy. The universe grants this gift of energy and asks only one thing in return: that we use it to say who we are. When we live a life that expresses our unique being, we align ourselves with the universe.

"I don't know that it's envy you feel. It sounds closer to cosmological dread. We shudder when we fear an anonymous sameness swallowing us. You fear Sagan has left you with nothing to do. That is to say, you think you two are the same, only he got to your task first. If you stay with that false assumption you will soon conclude you didn't have the right teachers, or the right parents, or that you were born too late, spiraling down into the ice of sameness Dante imagined as a vast funnel. I suppose a contemporary image would be a black hole sucking entire stars into itself."

"What's the way out?"

"Turn everything upside down. Admire what Sagan has done. Understand it as the contribution to world history that it is. Only when you come to a deep understanding of his work will you begin to see what he has left undone. That will be the arrow pointing to your own destiny. When that happens, you will see that you are not the same as Sagan. The two of you are part of a single, creative project."

"Wow," I said. "Wow."

"I've been through it," Thomas said.

"What do you mean? With Sagan?"

KEISHA BARGED through the swinging doors of the back room holding her pies in a white box. Thomas received her gift with both hands and they entered into what looked like a ritual, Thomas placing the bills next to the cash register, attempting to pay, Keisha scooping them up and forcing them back into his hand, then into his jacket pocket, both laughing. After some failures, Thomas dragged me in, claiming the money was mine, that I wanted to express my appreciation, that I'd be hurt if she didn't accept this. She flashed her radiant smile my way, then dashed behind the counter.

"Bagels," she said. She pushed the white bag to my chest. "You're Jewish, right? Am I right or am I right?"

It was too complicated to get into.

"Basically mixed," I said.

52.

A Cosmological Form of Intention

On the drive to the center, half a dozen ideas contended for my attention, but before I could choose, Thomas turned the conversation back to our neighbor, Margaret, who had fled from my poster of galaxies. Even though she was of course a complete stranger, he remained concerned.

"She would benefit by finding a way into what indigenous cultures call a 'friendship relationship' with the larger universe. The universe is not cold or even neutral. The universe *intends* Margaret's well-being. We in the industrial world eliminate a friendship relationship with all entities when we imagine the universe as a machine. But that has changed with the discovery of cosmogenesis."

"In what way has it changed?" I asked.

"The elegance of the universe's expansion is the way in which the universe expresses what might be called a cosmic form of intention."

"We're back to whether or not the Sun is generous."

"Yes. As you say, the crucial element here is that of intention, for that is how we identify acts of generosity. We need to understand that the universe can intend something even before human consciousness emerges. In the earliest moments of time, the universe was filled with chaotic interactions, violent collisions, and chance events, but even so, the universe found its way to construct complex spiral galaxies. There was no conscious mind intending a spiral galaxy, that is true. But intention lives at the level of matter itself. Matter is not inert or passive. Matter's intrinsic dynamism is the cosmic form of intention. As you can see, I'm using 'intention' with respect to energy, the same energy physicists and cosmologists study, the energy of the big bang. I am making the simple assertion that the ferocious energy at the beginning aimed at constructing stars and galaxies."

"In your thinking, did the universe aim at constructing Earth?"

"The early universe aimed at bringing forth the conditions that would make the emergence of rocky, Earth-like planets inevitable," he said.

"And that would hold for people too, right? You're saying the universe intends not specific people like Margaret. It intends the conditions for the emergence of people in general? Whereas in traditional religions like Christianity, there would be a sense that God wills even a very specific personality into existence, right? You're saying that in the universe story, we can't go that far. We can only say the universe intends the *possibility* for particular individuals."

"No," Thomas said, shaking his head. "The universe intends Margaret herself."

"But I just asked you if the universe intends specific people and you said it didn't." Thomas chuckled as if agreeing with me. But he was shaking his head as if disagreeing.

"This is another example of the inadequacy of English to express the universe story. The difficulty is the word *universe*, which in the industrial world is taken to mean everything 'out there.' If that's the meaning of *universe*, then it's true to say that the universe constructs not Margaret but a bundle of possibilities out of which Margaret will construct her unique life.

"But if we take *universe* to mean everything—including Margaret and her decisions—then it is true to say that the universe process intends

Margaret herself. To see this, we need to think of intensions as coming forth through time. The primordial fireball intends stars and galaxies; the stars and supernovas intend stellar systems with planets, and so on and so forth until we come to Margaret with her intentions, which are intentions within intentions within intentions. When Margaret decides to shape her energy into becoming a lawyer, the fireball and the galaxies and the living cells are all there, holding this region of the universe steady so that a new potency can be actualized. Thus it is that we can say, with full accuracy, that the universe intends to become a lawyer in the particular form called Margaret. Margaret's own intentions are the last in a long line of intentions going back to the beginning of time. The universe intends the unique Margaret who lives across the street from you."

"Then every decision we make is a universe decision?"

"Every decision we make is a universe decision because each decision is made possible by the long sequence of decisions that precedes it," he said. "That is why I prefer thinking in terms of dynamism. The dynamic energy at the beginning of time was *rushing* to give birth to stars and galaxies. The dynamic energy of early Earth was rushing to bring about the endless forms of life. Dynamism is primary; conscious life flows forth from it. The discovery of cosmogenesis forces this reversal upon our thinking."

"I love this," I said. "I really do, but I know what will come next. Even though I understand it right now, even though it's all clear in my mind, it will soon drop away again and I'll be left thinking of the universe in the same ways."

"That's always the case when entering a new world philosophy," he said. "The mind needs time to test each element of the vision in order to decide whether to enter fully."

"Can I speed up the process?"

"Certainly. By journeying with others. The evolution of humanity takes place in communities. I started the Riverdale Center to deal with the supreme challenge of transforming the human being. I wanted to provide a setting for reflection on cosmogenesis for my students at Fordham University. Over time, a remarkable group of scientists, artists, politicians, and religious leaders have joined the conversation. I don't know how long I can keep the Riverdale Center going, but it's functioning now and you're here. I tell you all this to give you a sense of what is coming. You might meet some people today who will become lifelong collaborators with you in the exploration and teaching of the universe story."

53.

Gottfried Leibniz at the Riverdale Center

Ten o'clock at night, Thomas's talk long finished, the dinners of chicken, mushrooms, and Caesar salad all consumed, a small remnant from the original crowd hanging on to the party atmosphere, and I am writing in my green, pocket-size notebook in one of the stalls of the men's washroom on the ground floor of Thomas's Riverdale Center. My mind became so packed with ideas and conversations I needed to retreat and capture some of it in language before it vanished forever. I put the lid down on the toilet, sat on it, and jotted from memory the insights from the roaring waterfall of stories flowing from the people Thomas had drawn together.

This practice of capturing the events and ideas of the day began years before when I was in graduate school. After being frustrated with myself for forgetting key thoughts that would come to me in the middle of a conversation—thoughts so important I was sure I would never forget them, but then forgetting them anyway—I learned several systems of memorization, including those employed by Gottfried Leibniz. It was Leibniz who, in a roundabout way, was responsible for my practice. A complete accident.

In graduate school, on a lark, I thought I'd investigate my own intellectual lineage as I settled into my research working with Professor Richard Barrar at the University of Oregon, wondering where he got his start and how he fit into the accumulation of mathematical ideas going back many centuries, beginning with his own advisor, Eric Hans Rothe. As I worked back through my mathematical ancestry, I was thrilled to learn my direct line included mathematical giants such as Lagrange, Klein, and Cauchy. But the big surprise came when I got to the end of the historical record in the seventeenth century and learned I descended from the father of the polymath Gottfried Leibniz, who had bequeathed a number of world-altering ideas, including the differential calculus, the binary number system of computer science, and the philosophy of cosmic harmony, which deeply influenced both Charles Sanders Peirce in the nineteenth century and Alfred North Whitehead in the twentieth. This was almost too much to believe. That Gottfried Leibniz and I had both learned mathematics from the same person, his father. Gottfried, while sitting in his dad's lap; myself, while studying with one of his father's intellectual offspring ten generations later.

As I pondered the meaning of this, my first thought was that it amounted to nothing at all. So what if tracing back from my advisor two and a half centuries I came to Leibniz's father? It's not as if this was a DNA connection. It's not as if I had received something unavailable to anyone with a library card. But as I moved more deeply into my research with Professor Barrar, I discovered an eerie truth every mathematician and physicist and geologist and chemist and biologist eventually learns. The scientific language that ends up in the journal article or scholarly book is not the whole of the understanding. Something is left out. Sometimes what is left out feels small, other times vast. The eerie feeling comes with the realization that what's absent can't be put into language, can't be described, can't be named. If it could be, it'd be included.

It was in studying Henri Poincaré's work on the solar system that I first learned this. The great French mathematician had found a form of mathematical chaos in the dynamics of the solar system, and in the midst of his mathematical article on this discovery he expressed his despair at ever giving an adequate linguistic model of it. I was frustrated and annoyed by this. Professor Barrar's office was on the fifth floor of Deady Hall where the Department of Mathematics at the University of Oregon was housed. I burst in without knocking. He was bent over his mathematics at his desk, the top branches of the maple tree visible through the window behind him. He was wearing his usual white shirt open at the collar, and he met my disrespectful intrusion with a smile. I had questions. What could Poincaré possibly mean? Why didn't he just tell us what he saw? Professor Barrar glanced at the passage, started to speak, interrupted himself, stood up, and went to the small blackboard on the wall. He wrote down the equation Poincaré started with, made some adjustments to it, got it into a form he liked. He smiled at me, brushed his short sandy hair with the palm of his hand, and explained in English what he had in mind. Then he turned back to the blackboard and developed the equations further. At times he would use his hands to help get the idea out. He drew graphs of homoclinic singularities, erased them, wrote down additional equations, waved his hands, spoke in half sentences.

The moment came, half an hour later or so, when I understood. It was one of the beautiful experiences of my life. In English, it was understanding how an infinite number of periodic orbits surround every homoclinic singularity in our solar system with its star and planets. I saw the impossibility of drawing it. Professor Barrar saw something in my eyes, or perhaps in my body posture, I don't know. He stopped writing. Smiling, his mouth slightly open, he nodded his head in affirmation of his pupil's progress. He could see that I saw. He could see that I now saw what Poincaré had seen that had led to his despair at ever representing what he saw in a mathematical graph. It was not just the mathematics that Professor Barrar had created and that I had studied. It was his person, his presence, that were also fundamental in awakening my mind. Just as his advisor had done to him. And back through history. If I came to understand anything, it was because the fire had passed from body to body through the entire lineage.

It was my realization that to understand something means much more than to understand "the linguistic symbols." In conversation and in reading, something takes place analogous to a spark. In the more intense moments, it is a flash of knowledge passing from one person to the next. They go together, the symbols on the page and the fire from the person. That's what I was after as I sat on the toilet lid, the symbols that would bring back the fire that had flared in contact with others.

After writing madly in the bathroom stall for several minutes, I heard the scrape of the door opening and I snapped my notebook shut as if guilty of a crime. My conscience was not the least troubled. In fact, I regarded what I was doing as one of the best things I could do. I had wondered more than once whether keeping track of the ideas that swirled out of Thomas, this rare human being, as well as from those closest to him, might become my primary gift to the world. Even so, even though I thought my self-assigned task had a touch of nobility to it, I needed to remember the context. Sitting on a toilet seat recalling snippets of conversation was not exactly what I had in mind when I imagined coming east to study with Thomas Berry. I'm not sure I would tell anyone the full story other than Denise.

My session over, I flushed the toilet to follow through on my pretense. I stood to leave and opened the gray stall door. An older man, upper sixties, in a dark green, short-sleeved shirt, was leaning against the wall. He had a pencil in hand and was jotting onto a three-by-five card. When I opened the stall door, his eyes flitted to the notebook in my hand.

"You too?" he said. His pudgy face broke into a huge smile.

"Pardon me?"

"Taking down his words. I've been doing it for years. You'd think I'd know by now, but he's always coming up with something new."

54.

Window to the Next Axial Age

Circling back through the living room, I entered the kitchen from its alternate door and found the liveliest group in the house, some washing the dishes, some drying, some putting away, still others jammed onto the bench behind the breakfast table. Everyone talking at the same time. A great bear of a man seated at the other side of the room pointed at me.

"He's right here," he said. "Ask him."

Mary Evelyn Tucker, a doctoral student from Columbia University researching the Neo-Confucian thought of Kaibara Ekken, "the Aristotle of Japan," was half seated on the windowsill. She was the first person I had met that afternoon. She had introduced herself with a warm smile when I sat down in the black plastic chair next to her, then turned back to her book. Out of the side of my eye, I could see she was reading something written in Chinese characters. Five minutes before the talk, she stood up, made her way to the front of the room, quieted everyone down, and introduced Thomas.

Now, in the noisy kitchen, she had a question for me, or rather her husband, John, did. Her red curls swung about as she searched for John, who had slipped out for his evening prayer smoke. As we made our way down the hallway to the back door, Mary Evelyn explained that Fr. Thomas, whom she adored, seemed overly pessimistic when he claimed that if we failed to reinvent the human species at this time, it would mean we'd lost the opportunity forever and would be forcing all future generations to live on a ruined planet. She understood Thomas's strategy of presenting his vision in a dramatic way in order to awaken people from lethargy, but she and John were wondering if he might be overdoing it. Did his dire warning make scientific sense?

John Grim was standing under the spreading red oak, taking in the setting Moon. He was stocky and muscular, a dreamer from North Dakota. The lights from the kitchen reflected off his glasses and teeth, which were bared by his smile. Mary Evelyn introduced him as a professor at Sarah Lawrence College where he taught courses exploring the shamanic traditions of the Plains and Northwest Indians. I jumped to my response.

"From the perspective of the macrocosm, what Thomas says is true. The major creative events in the universe are one-time and irreversible. So for instance, the universe brought forth galaxies in only one era. In the era before, the density of matter in the universe was too high for the existence of galaxies. In the era after, the matter was too thinned out for galaxy construction. There has not been a single galaxy constructed since that time, thirteen billion years ago. But when the conditions were right, a window of creativity opened up and allowed a trillion galaxies to flutter in.

"It's why I love Thomas's phrase, *a time-developmental universe*. The evolution of the universe is analogous to the development of a living entity, even an embryo. There is a sequence of these windows of creativity. The first window is for stabilizing the plasma into enduring matter, the second for constructing hydrogen and helium atoms, the third for the emergence of galaxies, and so on and so forth. It blows my mind to realize we live in something like a developing being with its own sense of creative unfolding. Although galaxies continue to collide and merge together, there was only one time when the universe could construct galaxies out of clouds of matter, just like there is only one time when a human embryo can create its eyes.

"Anyway," I said, "the question of whether or not this is true for events of human history is beyond me."

"What if we shrink the context?" John said. "Humanity as a whole might be too much to handle. But perhaps a one-time event is taking place in a subset of humanity, even right now. One that involves you."

"What do you mean?" I asked.

"Thomas has told us—I don't know how many times—that the universe story will never take hold until scientists embrace it. And here you are."

"Am I the first scientist to come here?"

"Pretty much," John said. He turned to face west.

"There's more to it," Mary Evelyn said. She looked at me. "I don't know how much I should say."

John stared westward a moment, then spoke.

"Just give him his medicine!"

Mary Evelyn waited a moment.

"When Father Thomas returned from his trip to Chicago at Matthew Fox's institute . . ." She looked at me in the dark.

"Yes?" I said.

"Several of us noticed a change. I asked him about the trip, but he didn't wish to speak of it. This was unusual. He always enjoyed regaling us with stories of his travels. I knew something had taken place. Not until weeks later did he tell me about meeting you."

"He senses the presence of a one-time event," John said. "One of your 'windows of creativity' is opening. All the elements are here now. Riverdale Center with its community of brilliant seekers has become the womb for a new vision. We don't have a name for it yet, but that will come."

The back door opened and the sounds from the kitchen flowed into the quiet night for a moment. It was Thomas Berry. He gave out one of his characteristic whoops of joy, saying, "The meeting of the century!" He placed on the glass table a wooden tray holding bowls, spoons, and a round container of ice cream. As he scooped out the ice cream, Mary Evelyn caught him up on our conversation concerning whether it was a certainty that if humanity failed there would be no second chance. Thomas listened carefully and waited until she was finished. With his legs crossed, ignoring his own bowl of ice cream, he leaned back in the iron chair and sighed.

"It's true, as you say, that I tend to emphasize the negative aspects of our moment, which is my way of alerting people to the gravity of the transformation we find ourselves in. Karl Jaspers's work is of some help here. He uses the phrase *axial age* to refer to the time twenty-five hundred years ago when some spectacular personalities came forth and established the spiritual forms of our major civilizations. Confucius, the Hebrew prophets, Lao-tzu, Pythagoras, Buddha. All of them in the same century. I believe we are in a similar moment. This new axial age will concern not just one region but the entire planet. A new form of mentality is emerging, one inaugurated by the discovery of cosmogenesis.

"Jaspers's insight has been developed further by the philosopher of consciousness Eric Voegelin, who concludes in his massive study of the history of civilization that such breakthroughs are not restricted to that moment twenty-five hundred years ago but take place in rare moments throughout the human journey. It is the nature of humanity to transform itself. Even though Voegelin does not mention our own time as one of these major transformations, he does offer his reflection on how they came about in the past. Through his careful linguistic, archeological, philosophical, and historical analysis, he discovered that the root of these transitions is human experience. Not philosophical ideas, not religious beliefs, not sacred scriptures. New human experiences led to the transformations of humanity. In our situation today, this would mean direct experience of a developing universe."

Vibrating with energy coursing through me, unable to remain quiet one moment longer, I interrupted him.

"That and that alone?" I asked.

Thomas was tired. He had slouched down in his chair, his hand supporting his cheek. Without moving his body, he glanced at me and nodded.

"But how do we make it start?" I asked.

"We don't need to. It's happening. But we can join this movement in a deeper way by reflecting on our experiences of the universe at work in us. The memories begin to surface as soon as we ask the question."

55.

Fraser River Awakening

When I awoke the next morning, I knew exactly which memory I wanted to reflect on. It was the summer before my last year as a doctoral student at the University of Oregon. Dad and I were on a fishing trip in British Columbia. While he dealt with the horses back in the barn, I sat on yellow sandy soil. The Fraser River was a mile downhill through manzanita bushes and sparse pine trees. In a lazy mood, I picked up a rock the size of my fist and dropped it. I repeated the action. My mind was filled with mathematics because of the concentration needed for dissertation work. At some point I became mesmerized by this action of picking up and dropping the rock.

The picking up part I could understand. My fingers clasped the rock and lifted it up. Nothing could be more normal or understandable. But when I loosened my grip, the rock *flew away* and landed with a thud on the sandy soil. The first dozen times I did this, nothing seemed strange about it. What could be more ordinary than a rock falling to the ground? But as I kept at it, and it fell over and over and over again, it became strange. Every time I released the rock, *something was taking it away from my hand.* Something was forcing it to land with a thud onto the ground. *Every time. Without fail.*

In that moment, I saw that my subconscious mind had tricked me into believing I knew what was going on. I had an explanation built into my perception of the event, which was, *things fall down.* Humans had lived with such an explanation for millennia. Even philosophers as sophisticated as Aristotle held the same belief. *But it is not even true to say that things fall down.* As everyone now knows, since science and technology have enabled us to leave the surface of Earth, if someone way out in space releases a rock, the rock *does not fall down.* Even if the rock weighs a ton. *Nothing will take it away from the hand that released it.* It will not fall away.

Even though I knew the scientific theories of gravitation inside and out, I still perceived the falling rock in the frame of mind that humans occupied many thousands of years ago. None of the theories had penetrated into, and changed, my actual perceptions in the universe. I was stunned to realize this. Even though I had spent years studying gravitation, I was experiencing the fall of objects the same way as ancient humans who knew *none* of this science. We both assumed the rock fell because *things fall down.* There had been no integration of my scientific knowledge with my life. My detailed knowledge was split off from my experience.

To try something different, I continued dropping the rock, but as I did so, I consciously brought to mind our best scientific understanding of gravity, which is Einstein's general theory of relativity. The great breakthrough of Einstein's work is his assertion that gravitational attraction comes not as an external law imposed on the universe but from the objects themselves. Though the mathematical equations are complex, the interpretation is straightforward. Space is imagined as malleable, and matter is pictured as having the power to bend, dent, and curve space. A two-dimensional analogy would be a vast plain made of a rubbery material upon which various objects like stars and galaxies rested. A single star would make a dent in the rubber surface, a single galaxy would make a deeper dent, and a cluster of galaxies would make an even deeper dent in this imagined surface. In this way each of these objects was a creator of gravity. Einstein's theory asserts that objects move along geodesic pathways that are determined by the curvature of this rubbery surface. If a rolling marble happens upon a dent in the surface, it will roll downhill toward whatever is causing the dent. If the marble happens to be moving quickly, it will slide toward the bottom of the dent but will have enough speed to carry it up and out of the indentation.

Applied to my situation there at the lip of the Fraser River in British Columbia, my rock was sliding down to Earth because of the dent Earth made in the rubbery fabric of four-dimensional space-time. This cosmological dynamic received a succinct summary by John Archibald Wheeler, one of the main developers of Einstein's theory, who said, "Matter tells space-time how to curve and curved space-time tells matter how to move."

The precision of prediction is astonishing. By plugging into Einstein's field equations the values for the mass of my rock and of Earth, one can predict with highest accuracy the pathway the rock travels when released. Einstein's work holds not only for the movements of rocks dropped on Earth, but for planets revolving around the Sun, for the Sun revolving around the Milky Way galaxy, for the Milky Way pinwheeling about Andromeda, and for the Virgo supercluster of galaxies soaring through space with respect to the other ten million superclusters that make up the universe. The predictions of all these movements can be made because their pathways are mathematical forms inside Einstein's equations.

What would it be like to perceive my dropped rock, not with the subconscious assumption that *things fall down*, but with Einstein's equations in mind? I returned to the image of a vast malleable plain upon which Earth is making a sizable dent. Now, when I dropped the rock, I imagined it sliding down the curvature of space-time until it was stopped by the ground. The rubbery plain was a conceptual apparatus for imagining space through Einstein's equations. I started imagining what it would be like if I were a thousand miles up from the surface of Earth. Earth's dent would be less steep, so the rock would slide out of my hand with less speed. But just as I was getting comfortable with viewing the rock from the insights of contemporary physics, a penetrating question hit me: Why does the rock slide *down* the curvature? Why not *up*? Why not *sideways*? Wasn't that the whole point, to *explain why the rock drops down*?

EINSTEIN'S GENIUS was to use the geometry of Riemann and Grassman to construct mathematical equations that were the exact representation of all macrocosmic motions in our universe. The predictions were even more accurate than Newton's theory of universal gravitation. The equations made predictions about the fall of the rock, but they did not provide an *explanation* for why the rock fell, other than to say it followed the curves of space-time. This was not an explanation of the movement so much as a mathematically precise description of it. By bringing Einstein's mathematics to mind, I hadn't changed the situation the least bit. The ancient assumption that "heavy things roll downhill" had been smuggled into Einstein's theory by the metaphor of a rubbery surface with marbles rolling into the dents and valleys. The question, "Why does a rock fall?" is identical to the question, "Why does a marble roll down space-time curvature?" Even though I had learned the advanced mathematics necessary to understand the general theory of relativity, I could not "explain" why a rock drops to Earth's surface. My bafflement was embarrassing. If anyone should be able to give a reason for why things roll down a hill, it would be a grad student who had put in the time to master the theories of gravitation. But that was *me*. That's what I had studied. I *did* understand the theories.

But the rock's existence was outside of the reach of the theory. I was not sure how to say it exactly, only that one way or another the rock had taken on a larger presence. I explained the situation by saying, "The dropped rock is more fundamental than the theory." As inadequate as that phrase was, it captured something of what I experienced, and I was certain Thomas would be able to explain it all better than I could.

56.

Bewitched by Theory at the Broadway Diner

Our conversation on my memory of dropping that rock took place at our favorite booth in the Broadway Diner. Thomas was reading my reflections with his eyes held in thin slits. The staff knew him well, whether water boys, maître d', or cooks. He didn't need to make his needs known. Each time a waitress, whether assigned to our table or not, happened by, she filled his cup without interrupting his reading. I was still the newcomer. As they filled his cup they glanced at me, their faces asking the question. When Thomas finished reading, he smiled and placed both hands on the manuscript.

"Your story captures the centrality of context in all attempts to understand. *Homo sapiens* have left behind our original ape context and have entered a profoundly new habitat, that of symbolic consciousness. There are new challenges in this habitat, the most formidable of which is our tendency to become so fascinated with our symbols we push aside life altogether in a pathological desire to live in the symbols.

"With your falling rock, you escaped this for a brief moment. In the years to come, I think you will remember that rock as the spark that led to your larger transformation." He squinted as he spoke, as if he were remembering something painful, or perhaps in actual pain. "This was the moment when you deconstructed your bewitchment with Einstein's equations and confronted the universe in act. Einstein's equations help us in our quest to understand reality, but his equations are not primary. No equation, no theory, no symbol is primary."

"Then what is primary?"

"Life itself. The actions of the universe. The rock falls to the ground and we say, 'It was gravity that attracted it,' which is certainly true. But that attraction was acting for billions of years before we named it *gravity.* The attraction is primary. We construct a symbolic representation of it in our mathematical theory. But then we reify these theories, which is an intellectual sickness."

"What do you mean, we reify these theories?" I asked.

"We forget that mathematical equations only *point* to the universe. This mistake is so common it has been addressed by major philosophers of our century, including both Whitehead and Wittgenstein. Whitehead gave this error a name, 'the fallacy of misplaced concreteness.' Wittgenstein referred to it as 'the bewitchment of our intelligence by means of our language.' I bring these two giants of thought up only to indicate how pervasive this defect is in modern thinking. A conceptual understanding of this error is the first step. But in order to escape this bedazzlement fully, one must undergo a process similar to your own, a process whereby one comes to feel in a direct way that the act of attraction is primary, the theory derivative."

"But wait," I said. "Equations are essential in that they show dimensions of the universe we would otherwise never know. Like the reality of antimatter. Or the details of stellar nucleosynthesis. That's why I love Pythagoras's theory that identifies mathematics with the foundational structure of the universe itself."

"The mathematical sciences coming out of the tradition of thought initiated by Pythagoras represent one of history's most significant transformations of human consciousness. As you say, the equations provide deep insights into the nature of reality. It's not surprising that we become so enthralled. But to identify reality with the mathematical equations leads to a reduced form of human consciousness.

"In the worst case, the symbols encase our minds completely, which amounts, as I say, to a sickness of the intellect. There are a variety of forms this takes, including such derangements as sexism and racism. All of them have in common a mind that has lost its capacity to see what is before it. In this reduced state, we see only our reified theories. When I mention Whitehead and Wittgenstein, you might get the sense that this is a technical, academic discussion, but it is closer to the truth to say it is a struggle with evil. Fascist leaders create theories of their enemies, stripping them of their humanity so soldiers can inflict pain and death, confident they are killing something inhuman. Industrial societies have also succumbed. By reifying Newton's theory of inert matter, we do not see the beautiful communities of life on this planet. Our diseased minds see 'natural resources' and we end up destroying with little awareness of what we are doing.

"In your own case, you might find yourself tempted to say, 'Einstein's equations cause the rock to fall.' But Einstein's equations are a sequence of symbols. No symbols cause a rock to fall."

"So we should say what, 'Earth attracted the rock'?" I asked.

"The rock and Earth together create the attraction, yes. Attraction is the universe *acting*."

"Acting?"

"Yes. The universe itself is the source of this attraction," he said.

"The universe is like a person?"

"No." With some irritation, he shook his head. "What I'm saying is not complex. Bring to mind all that the universe has assembled. That activity on the part of atoms and stars and galaxies didn't just happen. It was *brought about*. The rock and Earth are attracted to one another because they find themselves in a field of attraction that they—and the universe—give birth to."

"So attraction is the universe acting, that's all that can be said?"

"Yes," he said.

"That's what you mean by primary? The attraction just *is*, it's not something we can explain?"

"It can be explained," Thomas said.

"Then what is the explanation for why a rock falls?" I asked. "I have to admit feeling ridiculous asking you this question after working six years to earn a doctoral degree in gravitation."

Thomas chuckled softly. He pulled the top of the creamer and poured it into his cup of coffee.

"Bear in mind that science and cosmology are related though distinct investigations of the universe. Science for the most part focuses on explaining *how* things happen. Einstein's general theory of relativity is a profound explanation of gravity in that it provides an understanding of how things move in the macrocosm. It gives the precise mathematical form, the geometry, which determines the motion.

"But you are interested in answering the question, 'Why?' That is the approach taken in cosmology. The why of things relates to their development, relates to the future toward which they are developing."

"Yes," I said. "That is my question. Why does gravity pull the rock to the ground?"

"To create galaxies and stars. To create planets. Gravity exists to carry out the universe's aim of constructing communities of every sort. I call this community-building dynamic the third basic law of the universe. This amounts to a cosmological theory of gravity, one that complements the scientific theory. It is universal in that it pertains to everything, whether we are speaking of the attraction between Earth and the Sun or between two humans."

"Wow," I said. "You're saying this theory explains even our experience of love?"

Thomas smiled.

"You have it backward," he said. "Theories do not explain our experience. It's the other way around. Our experience explains our theories."

"Wait," I said. "What do you mean by that? What can you possibly mean when you say our experience explains our theories?"

"It's nothing esoteric. It's simply noting that our experience of wonder *leads* to our theories. Darwin's experience of the beauty of the life-forms in all their complexity led him to devote decades to the study of life. His theory of natural selection is one of the most profound insights into the nature of life in the entire history of humanity. But like all theories, it is a partial truth. It could not account for all of our experience of life. For instance, it had to be extended into the realm of genetics following the investigations of Gregor Mendel and so on throughout the twentieth century. That is to say, no theory gives a full accounting of our experience of life.

"On the other hand, our experience as living beings explains the existence of all our theories. Einstein's fascination with light and matter led to a lifetime of investigation, which eventuated in his special and general theories of relativity. It's a simple point I'm making. Experience of beauty comes first. Our theories follow."

MY MIND lit up. Things had fallen into place. Bits and pieces of my life came rushing to mind, such as my argument with Shel in the Boeing Amphitheater when I raved over experiencing the elegance of the early universe. That's what it was. Shel had reified the equations. I had too. Both of us had fused the equations with reality so that they became, for us, gravity itself. But as the light broke in, I understood the error. The equations were *not* sinews holding the universe together; the equations were symbols enabling us to *contemplate* the real sinews of the universe.

Equally stunning, the same gravitational force that was operating *then* suffuses us *now*. The force of attraction that built the galaxies was right here. I had only to release my pencil from one hand to the other in order to feel it. Experiencing the thud of the pencil as it hit my palm was to experience the primal attraction that had built the actual architecture of the actual universe. Literally. That power was right there when I opened my fingers and allowed it to take the pencil away. That's what I was experiencing, the living universe in the act of attraction.

I blurted out my question.

"Aren't you saying gravity and romantic attraction are the same thing? That gravity is a form of love?"

He winced.

"Our difficulty, as I've mentioned, is the impoverished form of many modern languages. Our mechanical worldview sequesters all power words like *intelligence*, *wisdom*, and *love* inside the human frame of reference. With our imaginations frozen in this way, it is difficult for us to see the cosmological origins of attraction. This whole line of thought might be a direction for you in the future."

"In what way?'" I asked.

"By helping us experience ourselves as modes of the universe. That is what the new story of cosmogenesis offers. This need for a deeper understanding of ourselves has been noted by some of the finest minds in the history of philosophy. Immanuel Kant in the eighteenth century praised scientific knowledge in the highest terms, but at the same time was keenly aware that even with the very best science, humanity will remain in need of a poetic vision of the whole. Kant's way of explaining this is interesting: 'For peace to reign on Earth, humans must evolve into new beings who have learned to see the whole first.' That's what's needed. A poetic vision showing us how to begin with the whole of things." He looked me in the eye. "Without doubt, a certain amount of rejection and even ridicule will accompany you as you endeavor to tell this story. No one should make the mistake of thinking cosmology is for the faint of heart."

My own heart was pounding. He had more to say, but I had lost my breath. The crucial fact of primordial attraction, primordial intelligence. It just was. With a simple aim, to move closer. None of that was known when we roared off that night. Like rocks falling to the sandy soil, like ignorant stars circling the Milky Way. Thoughts came pell-mell. To remember those events as nodes of the universe would change the past.

57.

The Nuclear Force in the Santa Cruz Mountains

The very next day, following Thomas Berry's advice, I began the task of remembering the memories of falling in love. I was filled with a sense of expectation. After listening to Thomas Berry for eleven months, I felt ready to train my knowledge of the universe on myself to see what it revealed about the experience of love. New insights might tumble out. It would be like the moment half a billion years ago when the trilobites invented the first eyes. That radically new visual experience must have blown their little minds. That's what it felt like. That I'd been given a new eye, time-developmental cosmology. Using it, perhaps I would see things never seen before.

After a breakfast of cornflakes and toast, Thomas Ian and I walked through the forest of maples, black walnuts, and mahogany trees down to the creek that flowed into the Atlantic Ocean thirty miles away. While his young imagination transformed sticks into sailing vessels that he released onto the lazy creek, I sat down next to a blue-black composite boulder. I imagined the retreating glaciers that had left this rock behind fifteen thousand years ago, a faint reminder of what once had been. The boulder was just a tiny dot next to the mile-high ice that had dominated the entire region, but it was a stubborn presence even so, not to be denied. The same with the memories that came back to me from fifteen years before. They too were stubborn and insistent as I leaned back against the rock and wrote.

I FIRST saw Denise Santi at the senior ball for her Forest Ridge Convent of the Sacred Heart, in Seattle. As I lived in Lakewood, an hour south, I knew no one at the evening event other than Julie Gamwell who had invited me. The ballroom of the Rainier Club was packed with dancing tuxedoes and formal gowns. At one point in the evening, Julie insisted I meet Denise since the two of us would be classmates at Santa Clara University in the fall. Julie pointed a finger across the sea of beautiful girls. "See her? The one with the big smile," she said. "See her?"

"Yes," I said, keeping the excitement out of my voice.

What does it really mean to say that I saw her? Photons of light had traversed the distance between us, bringing me into contact with this seventeen-year-old girl who was laughing hard, her head tilted back. An open-mouthed laugh. Did I see that she and I might have children someday? No. Did I see that the rest of my life, my destiny, would be shaped by her more than anyone else on the planet? No. Though such potentialities were there in the photons that arrived in my eyes, none of them surfaced in my consciousness. That is the way it always is with the universe. Five billion years before this dance, at a time when there was no Rainier Club, no Forest Ridge Convent, no Seattle, no Earth, and no Sun, there was a cloud of hydrogen and helium atoms. The atoms were attracted to each other, and in that attraction, and because of it, the Sun and Earth and Forest Ridge Convent came forth. Those potentialities were there in the initial attractions between the atoms, but the atoms had no awareness of these, just as I had no awareness of the ocean of potentialities longing to become actual.

Though I am pointing to an analogy between the attraction linking atoms together and the attraction connecting humans, I do not want to suggest the two are identical. The difference is at the level of consciousness. When Julie pointed out Denise, even though I was ignorant of the vast array of potentialities dangling in the moment, I certainly had more awareness than those atoms had back in the presolar cloud. If I had been asked to say what was going on in my mind, I would have spoken of my irritation at having to wear such stiff clothing; I would have admitted to my desire to impress Julie. But there was a dominant emotion that pushed all other contents to the side. As I was being pulled by a beautiful young woman through the dancing throng to meet yet another beautiful young woman, my awareness was dominated by the conviction that life as a seventeen-year-old was the greatest thing in the world. Which is a fine emotion to feel but one that lacks any sense of the depths of time involved. I was not just feeling an attraction; I was composed of attractions. If the elementary particles had not been attracted to one another to construct the light atoms, I would not exist. If the atoms had not been attracted to one another to construct the stars and galaxies, I would not exist. If the minerals of the rocky planets had not been attracted to one another to construct living cells, I would not exist. We live inside that nested sequence. At the base of our existence, we are an attraction inside an attraction inside an attraction.

JULIE'S DESIRE to introduce me to Denise went unfulfilled. As we pushed through the crowd, we kept bumping into her girlfriends who shouted to be heard over the harsh driving beat of the Sonics. I was sweating. Like a cloud of imploding atoms, I was growing hotter and hotter. My undershirt was soaked, drops of perspiration streaking my cheeks, but even before we were halfway through the crowd, Denise and her date had moved elsewhere. Attraction is the fundamental dynamic of the universe, but randomness is at work too. Chance is real. No one can predict the pathway of any particular atom.

SOPHOMORE YEAR, autumn, Santa Clara University, 1969. I was mesmerized by something Sherb, my physics lab partner, had said. He and I and

Jolly Spight had planned a trip to the ocean beach at Santa Cruz in order to get away from parents weekend. With Jolly still half asleep in bed, Sherb tells me that on his walk over to my room, he met three sets of parents, including the mother and father of Denise Santi.

"You did?" I said. "Where?"

He gave me a quizzical look.

"Graham 100."

"What were they like?"

"Who?"

"Denise's parents."

"I don't know," Sherb said. "Average looking, I guess. Normal. What do you mean, what were they like?"

"Would they still be there?"

Sherb looked over at Jolly, who was smiling. He looked back at me.

"How would I know?"

RUNNING DOWN the two flights of stairs, across Alviso, down Market Street, sprinting across the Alameda, then slowing down so I could catch my breath, I arrived just in time. Denise and her parents were walking through the arched gateway of the Graham dorms. Denise made the introductions. Mr. Santi was tall and thin, his hair still black, a handsome fiftysomething with an Italian look. Her mother had to be 100 percent Irish, wearing a pink blouse and a pink skirt, with a dark pink jacket. They both smiled warmly as we shook hands. Then there was nothing to say.

"So," Mrs. Santi said, breaking the silence. "Are you two dating?"

"No," Denise said.

"Oh," Mrs. Santi said. "I just thought . . ." She looked at her husband and stopped. The awkwardness deepened.

"Brian is on the basketball team," Denise said.

"I see," Mr. Santi said, nodding his head slowly as if thinking about something entirely different.

"He went to Bellarmine Prep. In Tacoma," she said.

More awkward silence.

WHEN I got back to my room, Sherb said I should ask her out.

"You can use my dog, uh, I mean my dodge," he said, making it a thousand times we'd heard that dumb joke.

"I'm not asking her out," I said.

"Then what was that all about?" he asked.

"What do you mean what was that all about?"

"You running over there to meet her parents," he said.

"It wasn't about anything."

FALLING IN love takes place at some deeper realm than consciousness. This is because love arises out of our bodily existence, and each body is connected through gravitation with every other body in the universe. I was irritated with Sherb because he could see I was fascinated, while I was still in the dark. But it's always been that way. A review of the fourteen billion years of universe activity leads to an alarming insight: The universe does not ask for permission when it decides to invade you and use you for its own creative purposes. Did it ask the hydrogen atoms if it could draw them into constructing a star? No. Did it ask the unicellular beings if it could knit them into giraffes and cockatoos? No. Just so, the universe had bonded me to Denise without getting anything like my consent. It charged forth and did its deed while I stood on the sidelines, half-asleep.

A WEEK later, I realized Sherb was right. This was ridiculous. Ask her out, for goodness' sake. I made up my mind to do it even though I didn't think my chances were great. The Italian Club had nominated her as their candidate for homecoming queen and plastered her face all over campus. She became known as the "Seattle Smile." Guys with Porsches dated her. I had the beat-up Ford Econoline van we'd driven down to Mexico City and back. And that had a dead battery. How was I going to fight through the crowd to ask my question?

I needed a friend to shore up my courage. After I talked Jolly into joining me, we marched across campus to her Graham 100 dorm. She was out celebrating her birthday at Farrell's ice cream parlor. We drove there and found her at a round table in the brightly lit room with a small group of friends eating banana splits and ice cream in glass dishes. The only person I knew was Phil Grosse who had been on the freshman basketball team. Their conversations died out as we approached. They stared in silence as if Jolly and I were mafia hitmen. I wished Denise a happy birthday and asked if they'd sung to her yet. They hadn't. I took hold of her elbows and lifted her to a standing position on her plastic chair. We sang the song and she smiled through the awkwardness of it all. At the end, I took her hand to help her down, but before I released her, I asked my question. The band Chicago was coming to town, would she like to go with me? "Yes, okay."

Her answer left me delirious.

The concert was three weeks off. To get the tickets, I borrowed from Sherb, Jolly, and Jim Hansen. I imagined the evening from every angle, thrilled that I had finally made it happen. A week before the night itself there was a knock on my door. Opening it, I found Phil Grosse and someone else I didn't recognize. He had come to tell me that there had been a terrible mistake. Denise's former boyfriend, now at Washington State University, was coming down the very weekend of the Chicago concert. It was to be a big surprise. Denise knew nothing about it. Phil was hoping I would call off my date with her. Even before I reflected, I said that of course I would, that I understood completely. I wanted to be the romantic hero, the one who denied himself for the good of another, who sacrificed his dream. That would impress her, wouldn't it?

But the next day a new question arose. Had she changed her mind about going out with me? After she had had time to think about it, did she regret saying yes? I had asked her in a rushed moment. She was embarrassed to be lifted up on that chair, I could feel it. Maybe that was too aggressive. Or maybe she felt bad about going out when a former boyfriend was interested enough to travel a thousand miles to see her.

My doubts blew away with a comment from Hugh Larkin as we sat at lunch in Benson Hall. Hugh was the most handsome boy at Bellarmine Prep, the running back on the football team, impressive enough in high school to be lured to Santa Clara on an athletic scholarship. Denise was eating lunch at a different table with Mike Olson who suffered from cerebral palsy. Mike's speech was difficult to understand; he wore a bib to avoid the humiliation of being fed by someone else in the student cafeteria, which meant food items fell on his lap and the floor around him. As Hugh watched Denise talk and laugh with Mike, he pointed out that she was exactly the same with everyone. What this means, he said, is that she will always be kind to everyone in equal amounts. Never to one person in a special way. As soon as he said it, I understood. It was hopeless. There was nothing special between us. Hugh knew all about girls and romance, and I just had to accept that I would never be more than one of her friends. I concluded the attraction was one-sided, and gave up.

When the early stars had dispersed their elements throughout the Milky Way galaxy, the next stage of the universe's evolution, life itself, was at hand. The difficulty was that the elements with the potency for life—the carbon, hydrogen, nitrogen, oxygen, phosphorous, and sulfur—were spread out over hundreds of trillions of miles. In order for them to give way to their chemical attractions so that the complex process leading to the first living cell could begin, they had to be brought into close proximity. The universe accomplished this by drawing the molecules together in small planets like Earth. These rocky planets began as a random mix of all the elements in a molten state. It was this molten state that enabled them to sort things out. Heavier elements like iron fought their way toward the center while lighter atoms like oxygen wormed toward the surface. As the elements rubbed against each other, they reacted in ways ranging from non-interaction to excited attraction. Each individual encounter would depend on chance, but because of the overall mixing of the elements, many results were inevitable. So, for instance, though it was impossible to predict the exact pathway of any particular aluminum atom, it was a certainty that, over billions of years of Earth's churning, oxygen, aluminum, and chromium would contact each other and give birth to ruby gemstones. This is true of every planet with convection currents. The six billion Earth-like rocky planets in the Milky Way galaxy are loaded with rubies.

The key factor in this chemical creativity is mutual attraction. If it is one-sided, nothing happens, as when oxygen brushes against neon. But when a deep resonance between both of them arises, the universe gives birth to a new synthesis. So with respect to Denise, there was only one question to ask: Did she feel any of the attraction I felt? Were we like oxygen and silicon that discovered a shared attraction out of which the Himalayas came forth?

THE END of sophomore spring quarter, eight months after our date had been canceled, I invited some friends to our second story room in McLaughlin Hall, including Tom Lunceford, Hugh Larkin, and my new girlfriend, Roberta. It was to celebrate the end of the academic quarter. As word got around, twenty more were drawn to join us and pushed their way in. The density of people grew to the point where everyone was in direct contact with several other bodies. Right when the room was a solid slab of human protoplasm, Jerry Zander showed up, steadying a keg of beer on his shoulder with one hand.

To free up space, someone placed the stereo speakers on the window ledges so the music blasted both in and out of the room. Conversation meant shouting. The break point came when Fr. Ted Mackin, our professor of moral theology, arrived. I don't know how long he was out there knocking. I learned of it when Jim Hansen, using his cupped hand for an ashtray, told me someone wanted to talk with me. When I opened the door, there he was, standing solidly in place out in the hallway. A Jesuit priest, he wore the standard uniform of black slacks, black short-sleeve shirt, and white Roman collar. I could see his lips move but could hear little until I shut the door and blocked out some of the roar.

"The university creates a series of rules," he said in his soft voice. "These rules establish the order necessary for students to study and learn. You are destroying that order, and because of that, I could have campus security here in a blink. But I'll leave it to you to answer the question. Are you about anything more than destruction?"

I apologized with real feeling, telling him I would close it all down.

OUR HIGH-DENSITY party expanded out of the room with a great deal of hollering as we spread out in several directions. Some of us were drawn to the crowd of students in the covered walkway to Nobili Hall where they had lined up for the dance. The girl at the entrance door taking the money struggled to keep up. The backup of bodies kept growing until someone at the front of the line broke through the shut door and the student mob flooded in. Rock music from the stage, screams, shouting, the black light glowing on teeth and eyes, people bouncing into each other, dancing, colliding, shouting. Jolly lifted me up from behind and shook me back and forth like a dog catching a rat in its teeth, and while I was still up in the air, Jim Hansen, his arm around Jane Higgenstine's neck, insisted I swig from the vodka bottle he had smuggled in.

It was in the middle of the dance that it happened. A circle of bodies had formed around Zander, who had dropped to the floor to do the Alligator, the guys clapping to encourage his outlaw behavior. That's when I saw Denise. Her powder-blue blouse with the high collar I knew so well, her dark blue eye shadow, her straight brown hair splayed over her shoulders.

She glanced over with a quick smile, a sparkle in her eyes, then looked away. That was all. But something happened. Her glance brought to mind her imminent departure. She was off to Rome for her junior year. I had known this for months, but only in that moment did it dawn on me that she was really leaving. That she would really be gone. Her voice would be gone. Once, standing next to her at Mass, I had been surprised by the strength of her singing, and by the feelings it ignited. She had looked up and smiled as she continued singing, her eyes mocking me for my sudden silence. Or pretending to accuse me of something. That's what she did. Made fun. She read letters from her fourteen-year-old brother with all the minute details of his soccer game—tried to read it anyway. Bending over laughing, crying tears of joy. Over nothing. Nothing but the teenage antics of her kid brother.

IN LARGE stars, when the temperature in the core reaches a couple billion degrees, two oxygen nuclei marry and become silicon and helium nuclei plus a shower of gamma rays. Though the star is the location of this marriage of two nuclei, it is not the cause of it. Their relationship is the cause. The resonance between them ignites the nuclear force, which fuses them together. The same with humans. Each feels the overpowering attraction that leads to a new bond in the universe, a bond that they, and they alone, have created. Its existence might confuse them. It might confuse others. It might lead to destruction of a previous order because it will not be contained by former structures.

Following my impulse, colliding with people as I pushed through to her, I told her we had to speak. Half an hour later, she and I were in my red van, heading south on Skyline Boulevard, tracing the ridge of the Santa Cruz Mountains. We had simply fled. I spoke to no one, not Jolly, not Lance, not even Roberta. We drove all night long, talking, talking, talking, talking. She and I had never been alone before. The proximity all by itself created an upwelling of feelings with a rush of language. It was the same when two nuclei ignite the nuclear force. An endless flood of gluons flows back and forth to make the strong nuclear attraction happen. We had no idea where we were going, only that we were going together, weaving through the forests of the Santa Cruz Mountains as we remembered the senior ball at Forest

Ridge and our love of the Puget Sound with its rain forests and the frost that swelled the ground in wintertime and the way the clouds hovered over the Olympics and grew pink in the early morning.

A wild happiness filled my world so completely I could not think. Especially not of the damage I had done. To Roberta. Our relationship was ruined, and she had done nothing. Nothing. The same for my status as a student. For the last two years my modus operandi was to focus on classes I liked and skip the uninteresting ones, which this term were electronics and partial differential equations. Typically this meant that I would study day and night through the entire week before finals to make up for my shoddy performance, but this time my mind was too fragmented to concentrate on the mathematics. I would flunk both classes, would trash my grade point average, would put an end to my scholarship money. This would be saying goodbye to the order of my life. To everything. And yet, none of those consequences made a dent in my mind. I had little understanding of what was happening. All I knew was that I was going to chase after her. If she was going to Europe, I was going to Europe. For the next two days, I slept in my van during daylight hours to avoid seeing anyone. All that mattered was making it to the night. Hoping she would again sneak away from her dorm room in Graham 100 and plunge back into our unending conversation, back into the bottomless upsurge of words that formed the foundation of a new world.

"DAD, WE'RE hungry." Thomas Ian was holding an olive-green bullfrog that blinked up at me. The frog was no longer struggling, as if resigned to being the captive of this five-year-old human.

"I've got a peanut butter sandwich for you," I said. "I don't think you should share it with your new friend. It might hurt him."

From my backpack I pulled out one of the cellophane-wrapped sandwiches and handed him a half. Placed on the ground, Mr. Frog immediately started to hop away, and Thomas Ian, holding the sandwich to his mouth with both hands, followed along behind.

As I watched him walk down the path to the creek, it was hard to think that his existence would not have come about if any of a hundred chance events had not taken place. But the truth is, the universe insists upon the existence of some things, while remaining happily unconcerned about the nonexistence of an infinite number of other things, any one of which might have been brought forth. That's what came to me in that glance. The real possibility that she would go off to Europe and be swept out of my life forever. The terror of that imaginary future was as profound as my attraction to her.

Yes, the existence of our son rested on uncountably many chance events. But that was not the whole story. In the moment I became aware of a fundamental branch point, I ran down the pathway that led to Denise. Whatever would come forth after that had for its base that conscious decision that she was my life. Thomas Ian did not come out of a purely random process. He came out of a decision that transformed all the events of our past from chance to necessity. They became necessary in that they were just what they had to be in order for us to embrace it all and make it our destiny.

I watched Thomas Ian toss Mr. Frog into the creek, landing with a belly flop in the middle, the water carrying him downstream. For a twinkling instant, I saw them both as flowing downriver from way back, all the way to the fusion of nuclei in stars. It wasn't just an idea in my mind. The actual oxygen atoms in the boy's body and the frog's body had once existed in billion-degree star fire. I had learned to do a cosmological analysis on my memories, and it had led to this moment, in real time, where I experienced part of an event from the perspective of the universe. That was the aim. Learning how to live moment by moment in a time-developmental universe, a universe in the process of becoming. Making this part of daily life.

Was this the essence of Thomas Berry's teaching for me?

58.

St. Augustine's Rome in Flames

On May 23, 1983, the Cathedral of St. John the Divine in New York City put on a conference, "The Spirituality of the Universe," featuring a dozen major thinkers, including Thomas Berry and Matthew Fox as keynotes. The scarlet brochure boasted this as the first conference featuring both of them. The whole affair was the brainchild of the dean of the cathedral, Jim Morton, nicknamed the "green dean" for his decision to break with tradition and put in the pulpit not just priests but ecologists too.

The day before the conference, the speakers were to meet one another in the hope that some common themes for the conference might emerge. I drove to Thomas's place, and we took his Nissan Maxima the rest of the way down the Henry Hudson Parkway.

WE PULLED up behind the line of cars waiting to be allowed entrance into the close of the cathedral. We inched forward, but when we reached the entrance booth, progress came to a halt. When it became clear that the car ahead of us was denied entrance, Thomas had to back up. This took some awkward maneuvering. After the flapping of arms and reversing onto the street by several cars behind us, the denied car sped off. The gatekeeper wore a long black raincoat, though it was not raining. He bent his ancient face with its full white beard toward us, holding a clipboard to check off names. When he saw who was driving, his face brightened.

"Father Berry," he said. "Dean Morton is hoping you will meet with him in Cathedral House."

He swept his arm in front of him as if inviting us onto a dance floor.

THE CATHEDRAL and its half dozen buildings take up four blocks of New York City. And though the early wooden houses that once formed the surrounding neighborhood had been replaced by modernist apartments and, farther south, New York's towering skyscrapers, the architecture of the cathedral close had maintained its medieval tone for over a hundred years. The same Gothic traditions used to build Chartres and Notre-Dame were brought to America, which enabled English stonemasons to carve and assemble the granite rocks floated down on barges from the mountains of Maine. Surrounded by stone walls, spires, flying buttresses, and half a dozen buildings, the worshippers heard none of the street sounds of New York. The grittiness of a modern city was replaced by grassy gardens, maple trees, and wandering peacocks.

Thomas knew his way around since becoming an honorary canon some years before. We followed the stone pathways, cut across a garden area, climbed a flight of stone stairs, made our way down another hallway, darker than the first, and found ourselves at the dean's office, the door wide open. Dean Jim Morton, with his wild eyebrows and penetrating blue eyes, looked like the leader of some ancient, esoteric tradition. He sat at his oak desk doing calligraphy with a pen made from a white bird feather. The bookshelves ran to the high ceiling and were filled with dark, hardcover volumes. A sliding ladder was waiting in place behind him. When he saw us at his doorstep, he ripped his Ben Franklin glasses from his head.

"Our great sage," he said. He blessed us, using his quill to make the sign of the cross. "What are you brooding over, Thomas? Come on, let's hear it." He turned his handsome features to me. "The man has never taken a break. I got him to watch a football game once. At one point, I saw a far-off look in his eyes so I asked him what was going on. Was he disturbed by the last play? Did the official make a bad call? Would you like to guess what he said?"

"No," I said.

"'I'm thinking about the downfall of Western civilization.' This was the *Super Bowl*. The Raiders were rolling but got slapped with a holding penalty. Now it's third and twenty and here's Thomas, oblivious. So come on now. Out with it."

Thomas chuckled softly and smiled. Jim continued.

"He looks innocent, I grant you. The kind, elderly gentleman with the shoddy jacket. When he talks, he's so quiet you think he's mumbling. You have to pay close attention. Then you hear his words. Then you get what he is saying, and it's . . . it's . . ."

Jim's face turned serious. He finished by banging the desk with his fist.

"*Cataclysmic.* For those who can hear, life splits into two. There's the 'before' part when all is well in heaven and on Earth, and there's the 'after' part when you take in what he's saying and suddenly see a world engulfed in madness. Which includes Christianity. Why does the Vatican let you go on like this? What's your secret?"

"Anonymity," Thomas said.

"That makes sense," Jim said, looking in my direction. "I keep telling him to write this stuff down, but I can see his point. Staying under the Vatican's radioactive radar might be the right way to go. Thomas, quit stalling. What's in that mind of yours today?"

"I was thinking of the author of Revelations," Thomas said.

"Here we go—" Jim said.

"As I often do when I come here. I was reflecting on his long years on the island of Patmos in the Mediterranean as he worked on his writings. He was the most significant precursor for Augustine's meditations on the fall of Rome. It is hard for us to fully appreciate just how deeply this destruction shocked the world. Rome had been free from attack for centuries; it was assumed to be utterly invincible. Then, in 476, the Visigoths penetrated Rome's walls and pillaged the city. One needs to bear in mind that this was not just another city. Rome was the political, spiritual, and commercial center of Western civilization. And now, after eight hundred years of protection from outside invasion, the entire city was put to the torch.

"Augustine brooded on this collapse from his perch on the shores of North Africa. Staring across the Mediterranean Sea, he wrote as if he could see the fires in the distance. He pondered the situation from multiple perspectives. What had caused this disaster? Was it the loss of faith in the Roman gods? Was it the rise of the Christian religion? The meaninglessness and despair of that time led many people to commit suicide. Life was seen as something painful and miserable, as something to be escaped from. His response was to write a new history of the universe using the time horizons and genealogies of the Bible. Though his *City of God* replaced Virgil's *Aeneid* as the fundamental story of the West as well as providing the foundation stone for the Christian Middle Ages, he himself saw none of this. For the rest of his life, Augustine witnessed only the ongoing collapse of the Roman civilization.

"Our situation today is the same, except that the decimation of life we are witnessing is a thousand times worse than what Augustine dealt with. What is burning now is the Earth Community. The new story we need today will not come from any one individual like Augustine but from a great number of people, scientists, poets, musicians, storytellers, and more. None of us is the author of this story. The universe itself is telling us how it developed into the structures we see. Our role is to learn to listen to the story the universe is telling us. It is a stupendous story. It will alter the course of history, of this there is little doubt. What is required now is moving the story from mathematical journals to the vernacular."

Dean Morton turned to me.

"This is where Brian the Wunderkind comes in."

"I assure you I'm anything but," I said.

"That's not what I hear from Thomas." He waggled his finger back and forth between us. "You two had better get your story straight. You need to be aware a couple theologians are gunning for you, Thomas. You too, Brian. Sorry about that. Guilt by association."

"What are you saying?" I asked.

"They're in the library now. Getting their scopes in place. Building bunkers. Did you really think they'd give up Christendom without a fight? Protect your weakest places. That's where they'll take aim."

59.

Cathedral of St. John the Divine

The next day, Denise, Mom, our two sons, and I drove down from Mt. Kisco to Manhattan. My mother, who had flown in from Seattle the day before, held Baby Sebastian in her arms in the back seat. Her high cheek bones, golden hair, and Norwegian blue eyes were replicated in Baby Sebastian, who smiled up at her, as if in gratitude for having a fair-skinned ally in this family filled with dark eyes, dark complexions, and dark hair. I had persuaded Mom to come to New York, as I sensed this conference would be a major milestone in our lives. Also, she was indirectly responsible. In her broad spectrum of intellectual interests that ranged from Carl Jung to Mariology to Arnold Toynbee, she had come upon the work of Matthew Fox, which she passed on to me, leading to my friendships with Matt and Thomas Berry. Denise and I had decided to bring both boys so they could be at my first short talk with Thomas Berry. Not that I would be at the podium with him, but I would be at the same event. The day's festivities began in the cathedral with Paul Winter's consort filling the vast interior with music.

Barefoot dancers started on the altar then flowed down the steps and ran and twirled up the aisles, all of them dressed in black or white robes with long flowing sleeves. Our early arrival at the Cathedral gave us our choice of pews so we settled into the very back with the idea that if either of our boys made a great deal of noise, we wouldn't be bothering the whole assembled group and could ferry them out without much notice. Then it became obvious that the event would not fill even half of the cathedral's enormous nave, so we gathered up our stroller, our bag of food, our children's books and quiet toys, and moved halfway toward the altar.

When the music finished and with the applause still echoing off the granite walls, Dean Morton, wearing a brilliant red chasuble, made his way onto the altar, stopping to say a word or two to musicians and dancers and pointing to people he recognized seated to the side of the altar. All this took several minutes. When he finally arrived at center stage and turned to the assembled, his radiant smile welcomed everyone before he said a word. His presence was so different from what I had become accustomed to while attending Mass in grade school. Our parish priest wore a mien of someone in shell shock, as if the crucifixion had taken place a day or two before. Whereas Dean Morton, with the confidence of a gifted political leader, lived in the excitement the Apostle John might have experienced when he rolled back the stone and found no body. After stating that our speaker, Thomas Berry, required little introduction, he took five full minutes to extol him as the grandfather of Earth spirituality. When he finished he pointed up to Thomas in the pulpit so far above the ground floor I had not noticed him.

Thomas began by suggesting we consider the cathedral an image of the universe. This is why a cathedral's altar should be placed in the east end, so worshippers face the direction of dawn, which for our ancestors was the location of the world's origin. St. John the Divine Cathedral was similar to the universe in that neither the universe nor the cathedral was finished. He called our attention to the muffled sounds coming from outside, which he identified as chisels of steel hammering on stone, a soft cacophony attached to the cathedral for more than a hundred years now. This ongoingness of creativity was true for the entire universe. That was the profound meaning of our discovery of cosmic development. Everything in the universe, as well as the universe as a whole, was in the process of building itself. Such was true for us as well. Each of us was unfinished. Each of us—as individuals and as members of humanity—was developing toward something beautiful, something surprising.

"It brings to mind a story passed down to us from the Middle Ages of Europe." Thomas's words reverberated in the dark air. It was difficult to hear with the echoes coming off the granite walls. He paused at the end of each sentence to let the vibrations die out before beginning the next sentence. "Three men were carrying stones. The first was asked, 'What are you doing?' He replied, 'I'm carrying a stone.' The second man was asked, 'What are you doing?' He replied, 'I'm providing for my family.' Finally, the third was asked the same question, 'What are you doing?' He answered, 'I'm building a cathedral.'

"We live in a new time. We ask these same questions of ourselves, but we can now provide a larger response. In the activities of our daily life, we perform specific tasks, yes. In our intentions, we aim to provide for our families, yes. But we are also involved with a work so vast we could not conceive it in former eras. We are Earth. We are a living, sentient planet. We are carrying further the creativity that brought forth the stars and galaxies. Whether we are fully conscious of this or not, it is happening. We live in a time that offers a significance only matched by the birth of life itself four billion years ago."

AS HAPPENED so often when I listened to Thomas Berry speak, I was instantly soaring with excitement. I pulled out the green notebook from my backpack so I could capture some of what he was saying. I wanted to reflect on how I myself would answer these questions from his ancient story. Our son Thomas pulled on my shirtsleeve. I bent down so he could whisper in my ear.

"Can we go outside?"

"I need to listen, Son," I said, and went back to writing.

After a moment, he pulled on my sleeve again.

"Can *I* go outside?" he asked.

"It's too . . . I don't know this neighborhood."

"I'm hungry," he said.

"How about some Cheerios?"

"You promised milkshakes."

"Yes, I know. When this is done, okay?"

That seemed to take care of it. It was questionable whether this was such a good idea, bringing the boys to a conference. All I really knew was that the event at the cathedral with Thomas Berry was going to have great importance and my little family should be there, be part of it. I hadn't thought it through from their point of view. How was a five-year-old supposed to understand a talk by Thomas Berry? And Baby Sebastian was still nursing; what was he supposed to get out of this?

What disturbed me when I reflected on my work as father of a five-year-old was my uncertainty over whether I was conveying a sense of the sacred to him. A sense of what we should be drawn toward, what we should avoid, what we should value highly even if it required pain and suffering. In my childhood, all such questions had answers that were worked into rituals then layered into us as children at St. Frances Cabrini parish. Chief among these was the Mass, which told in symbolic form the essential story of the world, that God had created all things, that humanity had fallen, that Christ, through his suffering and death, had restored humanity's relationship to God, that we were to go forth and do God's work.

Every element of the Mass had its significance. It didn't matter that kneeling hurt. We were taught to regard the desire to slouch back on the pew as a temptation. To fight against it. We learned that physical success in dignified kneeling meant moral success. To conform to the sacred dance of kneeling, standing, sitting, standing, kneeling again, meant rectitude. It meant one had identified with that which was right and good. The priest spoke the solemn words, bells were rung, and the holy event happened again as he lifted the host and the mystery of mysteries transformed it into divinity for the salvation of the world.

60.

The Reinvention of Sexual Attraction

I had no ritual that was comparable for our sons. We were in between cosmological stories. There were no rituals, but even to speak in terms of rituals fails to name our bereft condition. Mass was not a "ritual" when we were children. It was the essence of reality presented in an intensified form involving incense, music, art, history. It was the microcosm of the universe. It shaped us right down to the marrow of our bones.

Barbara Begay entered St. Frances Cabrini grade school when I was eight. The door opened in the middle of the school day and there she was, huge brown eyes, hands hidden in a white muff, her stylish blue coat buttoned up to her pageboy haircut. My best friend, Pat Kelly, was entranced with Barbara. Pat, having flunked first grade, was more experienced than I about these matters, and he immediately developed a plan. The first step was to become friends with her brother, Doug, who was two grades ahead of us. That would enable us, on some Saturday, to ride to her house near Gravelly

Lake, walk up to the door, and knock. When it opened we would say we were there to see Doug, and we'd hope to be invited in.

He revealed his plan when we were alone in our "fort," a platform of two-by-fours nailed onto the thick branches of an enormous fir tree in the field behind my house. Our original idea was that we would first lay down this platform and then build a small cabin on it. Even a two-story cabin. But the spot in the tree we chose was twenty feet above ground, and the work of lifting the lumber up piece by piece, plus the hammers, plus the unexpected difficulty of driving sixteen penny nails through the board and into a living branch convinced us to be satisfied with just the platform. It was our secret hideout. No one could see us, not even if they were directly below. In that seclusion, I was certain his idea for catching Barbara's attention made perfect sense.

On the day of, we started the three-mile bike ride to her home with a sense of easygoing adventure, but as we got close, I stopped talking. When we arrived, I stared in silence at the red brick house sheltered on both sides by tall Douglas firs. Apparently this is where she lived. Day after day. In this very house. Planning it out in our fort had been easy; now that it was actually happening, things changed.

Pat walked his bike up the concrete path to the front door. He turned and signaled with his hand that I was to follow him. He knocked and waited and then the door opened. Barbara, barefoot, with a pink top and short sleeves, stared blankly. I could not hear what Pat said to her. After a moment she disappeared and Pat turned back, waving his arm with new urgency that I was to come. He sat on his bike, waiting for me to move, which I did, but in the opposite direction, pedaling as fast as my Schwinn could take me. I would not be able to use the words, but on some level I had fallen in love.

The intensity of my adoration reached a new pitch in fifth grade when the seating arrangements placed Barbara and me side by side, she in row 4, I in row 5. I dealt with this new and dangerous situation by never looking in her direction. Then the day came when she lost a piece of jewelry. I knew it well. It consisted of the word *Barbara* captured in gold-plated metal twisted into cursive letters. She wore it pinned to her green sweater up near her collarbone. We got the information that it was missing from our teacher, who made the announcement and asked us to search for it. The entire class of forty-four milled about the room, looking.

I was the first to find it. It was on the floor behind her desk. I didn't pick it up. Similar to my inability to look at her, or talk with her, I was incapable of simply bending over and lifting up her jewelry. This was *Barbara*. This was something she wore every day. This was something that touched her *body*. Using the tip of my shoe, I edged it closer to the wooden slat upon which our desks were riveted. This would make it harder for anyone to find. It would be in a place no one else would know about. This was an act of thievery, but that didn't matter compared to the thrill of knowing it was there, a secret intimacy, so secret she wouldn't even know about it. Ten minutes went by. Then Gary Fitzpatrick found it. He picked it up, walked to the cloakroom where she was searching, and offered it to her. I observed her face light up. Saw her thank him. Several times. Speaking his name.

My fascination with Barbara collapsed the following year, when I was eleven. The momentous change took place when she placed her rump on the edge of her pew in a half-kneeling, half-sitting posture. The impact on my consciousness was immediate. I had been mesmerized for three straight years and had regarded anything she touched as numinous, and yet her slouched body doused my erotic interest in an instant. My conscious mind had not the slightest idea of what had happened. But on some deep level, my body knew that her slouch was a violation of the medieval cosmology that had been layered into my cells. When she first appeared in the doorway of our fourth-grade classroom, her presence had awakened a response coming from the long evolution of mammals. But those two hundred million years of mammalian sexual evolution had been matched, and even overridden, by the ritual enactment of a medieval European cosmology.

This is what Thomas meant when he spoke of reinventing the human at the species level. A cosmology was not just words. A cosmology was a symbolic process where human symbols sank into, and became part of, the genetic process that assembled human beings.

61.

The Spirituality of the Supernova

Thomas Berry presented the main talk of the conference from the pulpit, each word echoing off the stone walls of the cathedral. He was reflecting on the development of the universe and was using the phrase, *the spirituality of the universe*. He said he used the word *spirituality* to correct a deformation in modern consciousness that imagined the existence of a "physical universe." Such a conception no longer made sense. It belonged in the nineteenth century. In the twentieth, we discovered that the matter of this universe—the only matter we know of—constructs life. There is no such thing, then, as "lifeless matter." Matter, in its very structure and dynamism, generates life.

A similar statement can be said about spirituality. If there is any spiritual dimension in any human throughout history, then spirituality is one of the potencies of our universe, a potency laced into the elementary particles of the primordial plasma. Our discovery of cosmogenesis has led to our realization that matter is spiritual as well as physical. Thomas predicted that his phrase, *the spirituality of the Earth*, would soon be discarded as unnecessarily complicated. Matter will be understood to be spiritual. No qualifying adjective will be needed. The spirituality of a galaxy is the galaxy's intrinsic creativity. The most direct revelation of spirit is the creative synthesis that has given rise to every entity throughout time.

THOMAS'S PHRASE had ignited an argument between him and others the day before at our speakers' luncheon. Dean Morton's prediction that Thomas would be attacked came true when Bob Reynolds, a theologian at Fordham University, accused him of pantheism. A couple dozen presenters crowded around a long mahogany table in the library of Diocesan House for our lunch of stew and sourdough bread. Speaking from the head of the table, Dean Morton started the conversation when he asked with bright enthusiasm that we formulate a statement expressing our common faith. He began with his own proposal: "All of us believe in a loving God." This came under immediate attack. Why use *God*? Why not *Allah*? Why not *Yahweh*? Why not *divinity* to make it even more inclusive? What about *Tao*? What about *Great Spirit*? What about the hundreds of other names and conceptions for God?

Throughout these exchanges, which once or twice included harsh criticisms, Thomas sat quietly. When it became clear no phrase would work for everyone, Morton moved our discussion forward by asking each of us to present a synopsis of our presentations, signaling Thomas to begin. He was tentative, slow to start, as if still absorbed in the lingering disagreements. He detailed his talk by telling us it would highlight the four eras of cosmic evolution—the birth of the universe, stars and galaxies, living Earth, and human consciousness. Reynolds removed his dark sports jacket and black tie as Thomas spoke, several times glancing across at one of his colleagues with a "Can you believe this?" expression.

The instant Thomas finished, Reynolds spoke up.

"This is pantheism," he said.

Thomas shrugged his shoulders, turned his palms up.

"I am only relating the narrative of the universe that science has discovered."

"Yes," Reynolds said. "I get that, but you're calling it spirituality. Science has nothing to do with spirituality."

"In one sense, that is true," Thomas said. "Scientists have provided us with the data of cosmic evolution. What is required now is an adequate interpretation of the discoveries of science."

"Without ever mentioning Christ?"

"The promise of our cosmic story is to provide a common context for discussion and understanding. We make that possibility remote if we insist upon sectarian language."

"Sectarian language?" Reynolds looked at Dean Morton. "The last time I checked, this was an Episcopalian cathedral. Jesus Christ is the foundation of our tradition. Do you, or do you not, believe that?"

Thomas smiled and itched his eyebrow before responding.

"I've spent my life studying the world's religions. I believe in each of them. I believe in Christ, certainly, as I believe in Kaang of the San Bushmen, in Buddha, in Krishna, in the Great Spirit of the Indigenous peoples. Each of these cultures offers wisdom."

"Thank you for your confession of faith," Reynolds said. "Even better would be an admission that Christianity is your own personal foundation."

"It's not," Thomas said.

"Then what is?"

"The universe," he said in his quiet voice.

THOMAS IAN tapped my arm.

"I'm hungry, Daddy." I rummaged in our satchel until I came up with the cellophane bag crammed full of Cheerios. As soon as I handed it to him, he worked the flap open and started eating them one by one.

I turned back to listen. Thomas tugged on my shirtsleeve.

"What?" I asked.

He beckoned with his fingers, indicating I should bend down. He cupped his hand to whisper in my ear.

"Are Cheerios like nuts?"

"What do you mean?" I said.

"Do they grow on trees?"

"No, no. They're, uh, wheat. They take wheat and scrunch it up and bake it and that's the Cheerio."

I watched him, waiting for his face to relax, indicating we were done here and I could get back to Thomas Berry's talk. But no. He was still wearing the wrinkle on his forehead.

"Why do they have holes, then?"

ONLY BECAUSE I did not have an automatic response to his question did I wake up to what was happening. If there had been a cliché answer in my mind for why there are holes in Cheerios, I would have tossed that at him as well. With Thomas Berry's story from the Middle Ages in the air, I realized that same story had just happened. I had answered in the tradition of the man who said he was carrying a stone. He was not wrong in saying that, just as I was not wrong in saying the Cheerio was "wheat." But what a stale and one-dimensional answer. Obviously a Cheerio is wheat, but it is also energy. Energy from the Sun, even from the birth of the universe. In my unthinking reply, I was putting him in a strictly human world. If he had asked me *where* Cheerios came from, would I have said, "The yellow box in the kitchen"?

I LEANED over so I could whisper in his ear and not disturb anyone.

"SON, THE truth is, your Cheerio came from the stars. Stars a million times the size of Earth blew up and scattered their atoms that would eventually become that Cheerio. Every morning when you have Cheerios for breakfast, you're eating a star." I watched him, wondering how he would take this in. Maybe it would be his moment of cosmic amazement, as when I learned a tablespoon of white dwarf star weighed fifteen tons. Or maybe it would be nothing to him, nothing but words.

"DADDY, AT school they said God made the stars. But who made God?" He looked up at me as I stared down at him, waiting for something to come. The pause brought a look of alarm in his eyes. He was only five years old, but he was skilled at reading faces. Did he recognize the look in my eyes as an admission I had no answer? He squirmed in the pew, as if in physical pain. "It's *hard* to think about," he said.

JUST SIX years ago he did not exist. Not even as an embryo. And yet it was all happening again, the profound questions of metaphysics. They seem to be woven into our DNA as deeply as the instructions for building our lungs. To become human, it is as necessary to churn with those questions as it is to breathe air. As much as I wanted to provide an answer that would take away his pain of not knowing, any pretend answer that I offered would only dampen his passion to forge his own perspective out of existence.

THOMAS BERRY turned to the central idea of his talk: "In the fourteen-billion-year history of the universe, the event that carries the greatest spiritual significance of all is the supernova." He paused and waited. The words filled the entire cathedral with vibrations that complexified as they echoed off walls and folded back into themselves, and then damped down to make room for the next wave of sound to emerge from his mouth. I was electrified by his notion that a supernova was the greatest spiritual event of all. There was no easy way for me to fit such an idea into the story I had learned from my religion and my school and my family.

62.

Theories of Death at the Hudson River Palisades

I was a split person. As a child I had learned that the Mass was where the sacred lived. The Mass and prayer. And the sacraments. Also in actions of visiting people in prison, following the Ten Commandments, giving money to people in poverty. Those ideas and activities gave life its ultimate meaning. And then there was the world beyond all that, especially the world as discovered by science.

When I was nine, my family and I watched the televised images of the dark side of the Moon taken by the *Luna 3* probe. Since Earth holds the Moon in a tight gravitational grip, it is always the same side of the Moon that faces Earth, so no humans had ever seen the dark side. I sat on the couch enthralled. We were gazing on lunar mountains no human eyes had ever experienced. If there was a way I could become part of this adventure in deepening our knowledge of the universe, I would take it. But I had never been asked to think of this as being spiritual. Just the opposite. The basic idea was that all of it, including my amazement at the dwarf star and the dark side of the Moon, concerned nothing more than the "physical."

THE DAY before, on the drive back to the Riverdale Center after Thomas's argument with Bob Reynolds, I asked Thomas about Reynolds's reaction, and he said in reply that it was more complicated than it might appear. Reynolds had recently lost his wife and had taken a turn toward a more fundamentalist form of Christianity. He needed solace. He and his wife had been inseparable. Bob was struggling to commit to life again. As I listened, I wondered what Thomas's own understanding of death and afterlife were. It seemed too coarse to ask him outright, so I came up with something nearby.

"Can cosmogenesis offer the same solace that traditional forms of religion provide? Concerning death, I mean?"

Thomas kept his eyes on the road ahead of us. He squinted and said nothing for a moment. This was an expression I knew well. It meant he was embarrassed by my question.

"Certainly," he said.

"Is it hard to say how?"

"In the traditional cosmologies of both the West and Asia, the universe is understood as complete. As I've remarked upon before, Dante is a clear example. He places Earth in the center of nine celestial spheres, enclosed by the empyrean, the mind of God. These heavenly spheres are understood as having been constructed at the creation of the universe, and they will continue through time without foundational change. That is to say, the geometry of the universe as a whole remains the same. Beings in the universe can change, but the universe itself does not change.

"The meaning of human life for Dante is given by these unchanging structures. If one lives a good life, one rises up from Earth at one's death and lives in one of the celestial spheres. Such a view is sometimes called a two-story cosmology. Those of us shaped by such a story find it nearly impossible to escape the intuition that God is above in the eternal realm and we are down below in temporality.

"But precisely that orientation has to be relinquished if we are going to flourish in our time-developmental universe. All of the traditional cosmologies, including Dante's certainly, need a radically new understanding of time, eternity, and holiness. This is undoubtedly a bold statement that needs to be modified in various ways. Even so, it has to be stated. The cosmology of Aquinas and of Dante needs to be replaced. I describe the change as a cultural transition from a fixed cosmos, where the human agenda is the ultimate meaning, to a cosmogenesis, where the creativity of the universe is the fundamental meaning."

"But how does that help with death?" I said. I held my breath.

"Eternity is no longer understood as being up in the heavens. There is no 'up.' When a Norwegian looks up, it's the opposite direction from when an Australian looks up. Up and down are relative terms in an expanding universe."

"So how are we to understand eternity?"

"Ahead . . ."

"Ahead, like at the end of time?" In my excitement, I had stupidly cut him off mid-sentence. Thomas fell silent. I jabbered as a way to get him to return to his response: "That would put the eternal realm way, way off in the future. Wouldn't it? Maybe trillions of years. Is that how you think of it? A realm way off in the future?"

We had arrived at the Riverdale Center and were parked out front. He had turned off the motor. He probably wanted to rest after the long day, but I needed more. As usual, he could sense what I was thinking.

"Time for a quick drink?" he asked.

IT WAS late afternoon, just before sunset. We sat across from each other in the kitchen in the back of his house, the yellow surface of the built-in table pockmarked with wear over the years. The sunlight through the leaves of the maples high atop the Palisades across the Hudson River had the same golden hue as the Dewar's Scotch whisky Thomas poured into the two shot glasses.

"The challenges of life demand our full attention and concern, so I don't normally entertain questions about death, or life after death. My basic orientation is that death is an intrinsic dimension of life. I am certain the universe will take care of us in death just as it has in life," Thomas said. "But all of us end up reflecting on this question sometime or other and perhaps now is that time for you. I'll give you my theory of death, and then you can tell me yours. No, let's start with Teilhard's theory, which I believe is unique in the tradition of reflections on death and the afterlife primarily because he is thinking in the context of a developing universe. The vast majority of human reflection on death, including that of Thomas Aquinas and Dante, takes place in the context of a fixed cosmos.

"The traditional question is whether or not a person survives death. Teilhard's answer is yes, and in that sense he stays within the Christian tradition, but his manner of reasoning is where he shows his unique vision. For Teilhard, development through time is the primary revelation. It is the fundamental source of meaning in the universe. By development he means the cosmic and organic evolution as discovered by scientists, but he includes his conviction that the process of evolution is entwined with the process of love, an idea he attempts to capture in his neologism, 'amorization.'

"Teilhard's thinking is that a complete annihilation at death cannot be the case because in order for humans to embrace the evolutionary challenges, they must have the sense that there is a way forward, that the future is open. If humans came to regard death as their end, they could still find value in caring for their families and others in need, certainly, but it would be nothing like what they would experience were they convinced their actions had eternal significance. In his later years, Teilhard's deep concern became the activation of energy. He saw nihilism not as a moral mistake but as a cosmological dead end. His primary objection to the notion that the universe is meaningless is that such a conviction enervates humanity.

"There you have it. Teilhard's faith in the universe's development leads to his sense of immortality. Teilhard felt humanity as a whole will one day achieve a deep conviction of immortality and this will be on the order of a major evolutionary achievement, along the lines of aerobic respiration or photosynthesis. It will lead to a massive influx of energy into the human adventure.

"As for myself," Thomas said, "my thinking is darker, not in the sense of cynicism or depression, but in the sense of an appreciation for that which lies beyond language. I don't believe that at our stage of development we humans have the cognitive capacities for understanding the deepest dynamics at work in the universe. Perhaps we will someday, but at the present time, the complexity of the universe far outstrips our theories. I like to say that to assume the complexity of the universe is captured in our theories is the same as believing Beethoven's symphonies can be rendered by beating a garbage can with a stick.

"In any event, my own personal orientation concerning these ultimate questions comes from the Palisade cliffs, which I gaze upon every morning when I write, through all the seasons, day after day, winter giving way to spring and on and on. We're companions now. Earth's tectonics constructed them, and they've been sending their presence in my direction for two hundred million years. I entered the conversation as Thomas Berry only a few decades ago. We don't exchange words, but even so, a communion takes place. Humans have expressed their faith in a great variety of symbols, many of which have inspired me at one time or another. But today, if you ask for the foundations of my faith, I would say the stone cliffs of the Hudson River Palisades."

63.

Cosmological Love

Thomas, up in the pulpit, was near the end of his talk. It was then that Matt Fox slid down the pew next to me. He had flown in from Atlanta where he had given a weekend workshop. Zipping open his black travel bag, he pulled out a notebook and pen. With his big blue eyes flashing behind silver-rimmed glasses, he whispered, "What I'd miss?"

"Thomas Berry getting accused of pantheism."

"What a surprise," he said. His voice was full of mirth. "Someone was paying attention!"

SPEAKING IN somber tones, Thomas declared that in the history of human culture, there have been many different forms of revelation, by which he meant experiences that become foundational for a person's life, as well as for an entire civilization. In Western religions, the most widely known form of revelation would be biblical scriptures, which are an illustration of human language as revelatory. When prophets state, "Thus saith the Lord," they know the words they speak come from themselves; but they also know there is more going on, that the language and the story it tells have a meaning beyond the personal.

"What we have largely forgotten," Thomas proclaimed, "is the most fundamental mode of revelation, the cosmological. The universe, along with planet Earth, both in themselves and in their evolutionary emergence, constitute the primary revelation of that ultimate mystery whence all things emerge into being. The universe's revelation is primordial.

"The most spectacular unveiling since the birth of the universe is the supernova explosion. In the twentieth century we have learned that a chemical alchemy takes place in the core of every star. The atoms of carbon are created by stars and poured out into the Milky Way. The creativity of stars is the one and only way carbon is constructed in the universe, which means that each carbon atom in our bodies, without exception, came from a star. There were no carbon atoms in the primordial flaring forth of the beginning of time. Only through stellar alchemy could carbon, with all of its potencies, appear. Humans flower forth from the supernova explosion like roses from a rosebush. We need to relate to this release of elements as we would to a gift. A gift that enabled life. Did the universe ask us to pay for this? No. Have we done anything that merits this cosmic gift? No.

"Traditional language refers to unmerited love as grace, as love freely given. Stars, then, are bestowers of grace. They are bestowers of life. This is true in a general sense, but we need to remember it was specific stars that fashioned the particular atoms of our body. Soon the scientists will trace our personal biographies back to the exact stars that gave us our bodies. Following that, the poets will give these ancestors names. We are their offspring. We are those three or four stars in a later form.

"Our frozen imaginations struggle to see stars as bestowers of grace because we have convinced ourselves they are objects. While it is true that our ancestral stars did not know they were giving birth to us, it is wrong to say stars do not know. They *do* know. They know how to create carbon, silver, boron, and calcium. They know how to participate in the ongoing development of the universe. They know how to fulfill their role in this spectacular process.

"The central revelation of the supernova is its irreversible gift-giving. Irreversible because the star uses its energy to fashion the elements, and once that energy is used, it is not restored. The gift requires the star's death. Though it is a one-time endowment from the star, it is an ongoing gift-giving from the universe. Scientists estimate that with the passing of each second, another star has exploded and is disbursing its treasures. This extravagant gift-giving is the spirituality of the universe. It is a form of cosmic love that enables the future to emerge. Our ancient epics extol humans who give their lives for the well-being of the community. Even if these authors knew nothing of supernovas, they were intuitively aware that the universe values generosity. The generous personality is the human mode of a supernova's extravagant gift-giving. What I have to offer in terms of faith is simple in the extreme. My trust is in the star's bestowal of grace."

64.

The Solar System Floats Light as a Feather

Even though I remained aware that I was sitting between Matt and our son on a pew, I was floating. Not that I was floating up from the pews. The pews were floating too. Everything was floating: the people and Thomas Berry in his pulpit and the cathedral with its massive walls and stained glass windows, all floating. I had expanded to become all of that. The granite walls were still the granite walls, but we were now all together as one floating thing. The container in which I had been living had fallen away and I became immense. It brought to mind the white bubble over the swimming pool at the Lakewood Racquet Club where I worked as a lifeguard in high school; Dad would turn on the air compressor and the white canvas cover lying on the surface of the water would explode out to its full size. Same thing, here in New York City. We were all blown up to our full size. We were the cathedral with its rock walls composed of oxygen, sodium, aluminum, silicon. All of us had once floated in a vast cloud with our bodies all intermixed, then sorted out by the creativity of time to become that granite pillar over there and that stained glass window up above. We were floating then, and we are floating now.

This feeling would soon collapse, but in the moment, I knew what was happening. I had identified with the whole of things, with the whole cathedral, and the whole Earth. I was experiencing existence from a larger perspective. For billions of years, the Sun and planets have floated light as a feather. There is no up or down for the solar system. These feelings would evaporate in an instant and I'd be my ordinary self again. But right now, sitting in the pew, I floated with the Sun and planets and watched Matt scribble madly in his notebook to capture Thomas's wisdom. He was always learning. That was the spirituality of his religious order, the Dominicans, Thomas Aquinas, the love of reading, thinking. Over a lifetime. We learned Latin in order to feel our way into it. The supernova had released silicon and oxygen, some of which became these walls and some of which became us. Matt had translated Aquinas into contemporary English, which became me, breaking out of the shell, became the pew, became Aquinas in English, became this floating experience.

I felt my heart beat in my chest. Pulse, pulse, pulse. A beating heart from the supernova's energies, I was this same supernova explosion, not that "I" was having this experience while surrounded by other I's who were not. The event encompassed us all, Denise, Mom, our children, the cathedral walls, Thomas Berry's words, the exploding star, my childhood. We happened together with the granite pillars and the Milky Way. Thomas described his life in the monastery when the monks would pray the Night Office at two in the morning and then go back to sleep until the Dawn Prayer at sunrise. But he never returned to his bed. Stayed up through the night and read and reflected. When his health broke, he kept on. That wisdom lived coiled in his words, in the tone and timbre of his speech. It was that which ignited our expansion. He ruined his lungs, could not make it through a talk without collapsing into severe, pulmonary coughing, which too was the pulse. Like a supernova, he had no idea how his bestowal might change things. Nor did he need to. He too had already died and lived only to give away the wisdom he had accrued from decades of contemplation, flowing out of him now, the pulse of the supernova up there in the pulpit.

The origin was here. Light from the beginning, the cosmic background radiation, suffusing the cathedral, bathing everything. A silky, ethereal fluid too subtle to be felt as it brushed against our skin. Einstein had shocked the world of science by announcing that time was variable, that time slowed down with velocity. There was no universal time but trillions of unique times, one for every entity in the expanding universe. Nothing could be more counterintuitive, and yet when physicists measured elementary particles moving at high speeds, the calculations of Einstein proved to be exact. Time slowed down, and the ultimate surprise was that at the speed of light, time ceased. This cosmic microwave light from the beginning of time was everywhere present in the cathedral and had not aged one second in its journey from the origin of the universe. Its arrival was the same moment as its departure. The origin was here, every moment of time in between was here, the temporal had become the eternal, wedded together.

Denise nursed her baby hidden under a white cotton blanket. Just a moment earlier, maybe two, it was my mother who was nursing, and it was I who was under the blanket. The profound joy in her eyes when she first became pregnant with Thomas five years ago. It was everything she wanted. The dream of it bubbling out that first night in the Santa Cruz Mountains, proclaiming she wanted six children, laughing with her erotic energy that reached for fulfillment. But though I too was excited, and though I wanted nothing more than to become a good father for this unseen child, my own joy was laced with fear. The fear of a fatal birth process. The anxiety that arose as I felt the vulnerability of infant, the breakability of mother. She had given her life away. Her flesh, her blood, her milk, her nights, her mornings, sacrificed to this universe in the form of a child.

In its belly, the star brings forth the elements. These are not simply in the star. They *are* the star. The gestation is long, tumultuous, and yet because of it, and it alone, life-generating elements of carbon, nitrogen, and phosphorous come into existence. All of a sudden, even if billions of years later, it's over. Like a salmon gutted for its eggs, the star is torn asunder. Dismembered. The pulse had become something new. Thomas Berry, Denise, Mom. In this quiet, all was clear. Here was the spirit I was in search of. Erotic attraction that led to life-ending sacrifice. That and that alone at the heart of the universe.

65.

The Universe Is a Green Dragon

Just after ten in the morning on May 24, 1983, the day after the conference, and one week before our planned drive across the continent to the West Coast, I set out to capture what I had learned with Thomas Berry. The dishes were stacked up on the kitchen counter and filled the sink as well. The pot was buried under a stack of plates. I pulled out the top two dinner plates, but there was nowhere on the counter to put them so I gave up on making hot oatmeal. I didn't want to clank the dishes and awaken Denise. She and I had taken turns all night dealing with our son, who was colicky. I washed a red plastic bowl and settled for Cheerios, milk, and a cup of coffee. With my green notebook next to the bowl, I quickly sketched out what I wanted to write about.

I was in a different universe, on the other side of a divide. Driving to the cathedral for the conference, I was confident in my understanding of Thomas Berry. Overall. Maybe there were details I needed and an additional idea or two, but I had read all of his mimeographed essays and I believed I understood them in the main. Without question, I could have taught a college course on his thought. But what happened in the cathedral listening to him changed me. It started off as a stroll on a sandy beach with pine trees and squawking seagulls, then an ocean wave swelled up out of nowhere and swept me out to sea.

The magnitude of the transformation had escaped me. Thomas had indicated as much in our first meeting at the Broadway Diner. He explained that taking in a new cosmology amounts to becoming a new person. It involves a change in the structures of our existence. I now knew what he meant. My experience listening to him would not fit into my former self. I had the same mathematical knowledge of the supernova explosion, but something other than the knowledge, something at the feeling level, had flowed in, unnoticed until yesterday.

WITH MY bowl of Cheerios and my coffee cup, I sat down at the desk to write. When we first arrived in New York, I asked our landlord if there was any chance I could perhaps write in his garage. He said he used it for one of his old cars. I said maybe I could find enough room off to the side. He shrugged and the two of us checked it out. He was right, there was no room. In addition to the car, there were boxes of old tools and assorted pieces of lumber stacked up against the walls. The one window, at the far end, was covered with soot and cobwebs. It looked as if no one had been there in twenty years. I asked if we could park the car outside, but he didn't want to leave it in the rain and snow. As a last-ditch effort, I asked if I could use some of the boards and bricks to build a makeshift desk in our bedroom. He agreed. I hauled a dozen cinder blocks and an old door up the staircase to our second-floor apartment. There was just enough room between our bed and the wall. Even the door's hole, once housing a doorknob, turned out to be the right size for my coffee cup. I was set up for writing.

Feeling the closeness of our departure from New York, I wrote nonstop for two straight days, getting up only to go to the bathroom or to make more coffee. Denise brought me food, which I ate at the desk. Whenever I was too exhausted to keep at it, I put my head on my manuscript pages and slept for a bit. I rushed to get it all down while it was bursting with energy inside me. As I wrote, my overriding feeling was a sense of obligation. I had been given the gift of this year from Bruce to learn the new cosmology from Thomas Berry, and now that I had it in me, I wanted nothing more than to send it out to others who might find their own pathway illuminated by Thomas's wisdom. Even though the task seemed impossible, I had to give it my all.

My plan was to recreate my experience of listening to Thomas, especially those moments when we sat under the red oak while he answered my many questions concerning the meaning of the universe, all of which I tried to capture in my notebook. I wanted to write a dialogue along the lines of Plato's *Timaeus* where Socrates and his friends lay out their story of the universe's construction. To employ the genre of Plato's dialogues was a bold move, aggressive, perhaps even pretentious with its suggestion that Thomas Berry was on a level with Socrates. But as I wrote, I asked myself, multiple times, "What gives a better account of the universe? Plato's *Timaeus* written from the perspective of a static cosmos? Or Thomas's "New Cosmic Story," written from the perspective of cosmogenesis?

The answer returned each time I asked the question: "Both."

IN THE early morning of May 26, I put my hands on the top of the stack of pages. I was done. After so many gallons of coffee, my whole body vibrated with the caffeine and my stomach hurt so much I wondered if I had given myself an ulcer. Thinking I could placate the pain with food, I buttered a dozen pieces of bread and consumed the slices one after another, washing them down with a quart of milk, which I drank right from the carton.

The next move was to mail the manuscript to Thomas. It was too awkward for me to hand the pages to him personally. What if he regarded my request to read it as just another burden? Or what if he started reading it in front of me, which might lead to my seeing displeasure on his face, even if he tried to shield that from me?

No. None of that. Mailing it was the best option.

The post office wouldn't open for a couple hours, which gave me enough time to make a photocopy of the pages and to come up with one last, important item. Throughout the two days of writing, I kept a separate sheet for recording ideas for the title. There were over two hundred of them. Most were takeoffs from the titles used by the writers I most admired. If I were Plato, I would call the dialogue *Thomas Berry*, but the possibility that he would reject such a title because he did not want to be identified with what I had written would be more disappointment than I could endure. My fear of having him say I had misrepresented him was so great I wouldn't even use his full name in the text itself, designating him as "Thomas" and myself as "Youth." To pull this whole thing off, I needed a title that would impress him, a title that employed classic, philosophical words along the lines of *Ontology, Quantum Physics, Phenomenology, Mathematical Cosmology, Creativity, The Initial Singularity of Space-Time.* A powerful title that announced the seriousness of the weighty matters between the covers of the book.

This was nothing more than treading water. It had already been decided. The title that screamed out to me was the last title idea after two hundred attempts. It was last precisely because the moment it appeared, I knew it was the one. *The Universe Is a Green Dragon.* Which I knew he wouldn't like. But maybe he could learn to put up with it. One thing was for sure, it came from some place deep inside, for the instant I saw it, I knew my search was over.

66.

The First Hexagram of the *I Ching*

On Wednesday, June 1, 1983, Mom, Denise, our two boys, and I started our trek back to the Pacific Northwest. This would be our last day in New York, which fact I kept bringing to mind in order to shake my foul mood. We had spent the last week packing up our belongings, hauling the cardboard boxes down to the Mt. Kisco post office, and cleaning up our apartment. After inspecting our work, our landlord decided he wouldn't return our security deposit, which was a blow because of our lack of funds, but my sour mood was not about that. It had been an entire week since I'd sent the manuscript off, and I still hadn't heard from Thomas.

Denise tried to convince me that he probably didn't have time to read it, and I was sure she was right, but as reasonable as this was, it did not make a dent in my dread. It was obvious I had placed too much importance in receiving his opinion, and that I was demanding, in silence, that his opinion be extravagant praise. I had convinced myself that nothing less would give the sense it had all been worth it. The departure, the years away. Invading my anxiety was Dante's image for how the high artificer dealt with a mind too fluffed up with expectations: a huge rock placed on the person's shoulders. That's what I was feeling, and its weight grew with each passing hour.

The plan for our last day was to drive to Riverdale Center, pick up Thomas, and go to the Metropolitan Museum of Art together. This was Thomas's idea, his way of saying goodbye. When we got there, he was standing on the porch waiting, just as he had been that first day. He piled in, sitting in the front with Thomas Ian and me. Mom, Denise, and Baby Sebastian were in the back. I thought—even hoped—he might have waited until this moment to tell me what he thought of my writing. But no. He focused the conversation entirely on Mom, calling her "Momma Swimme," asking about her childhood in Seattle and her work as a teacher of high school Spanish. Her replies were reserved and short. I could see she was deeply moved to be in his presence. He had brought a going-away gift for young Thomas, an arrowhead, which he called a "point," found on the grounds of Riverdale Center years before.

His ever-present graciousness, massive intellect, radical philosophy, and pervasive kindness eased me out of the misery that came from my demands. Just to be in his presence was to feel liberated. It no longer mattered that he had not read my work. This was to be our last day together. I was glad to be alive, glad to be with him these few, concluding hours.

When we got to the Met, he led the way to a new exhibit he thought Momma Swimme would enjoy, the Aster Chinese Garden Court, a recreation of life in the Ming Dynasty. He pointed out various features to Mom, referring to his own time in China when he was doing research into what he called "the most cosmological of all the classical civilizations." His erudition and his calm cadence led museumgoers to conclude he was a docent. Very quickly a dozen people joined us to listen to him. When we got to the hall holding the Chinese bronzes, he turned to me and said there was one he especially wanted me to see from the Shang Dynasty in ancient China.

He had me bend down close to the glass to get a good look at the bronze while he described it. The vessel had various black lines cut into it. As happened so often when I listened to Thomas, his words mesmerized me, altering my perceptions. The effect was subtle, difficult to describe. For instance, the bronze he singled out had the same colors of many oxidized bronzes, a light green speckled with the tiniest white spots. This oxidized patina gave the ancient metal a spongey look. Even though the bronze remained the same, my apprehension of it changed as Thomas explained that its symbology was connected to the *I Ching*, a central work of Chinese cosmology. This bronze, in particular, celebrated the creativity of the cosmos.

Staring at the foamy green bronze, knowing it was three thousand years old, learning that the artisan was celebrating the creative principle of the cosmos—all of this taken together led to an experience of the whole process. I became fascinated by the simple idea that at one time this bronze did not exist; then it came forth, erupting into existence. I imagined it happening right now, stunned by the fact that it had actually emerged out of Earth, that an ancient Chinese artist was doing exactly what I wanted to do. His medium was bronze, mine was language. Both of us focused on awakening an experience of cosmic creativity. It was not just the bronze that had materialized; the Sun, the Moon, the oceans, all of them had surfaced into existence. As I stared, Thomas directed me to look from a slightly different angle. He pointed to one of the black designs. He said it was a symbol for a nose. The two smaller markings near the upper rim represented eyes. The Chinese artist had chosen to celebrate the first hexagram of the *I Ching*, the creative principle, with the face of a mythological creature. While I continued staring, Thomas gave me one last bit of information.

"Mythologically, the first hexagram is depicted by the dragon," he said.

It took me a moment. Items of thought and sensation were floating about my mind. Then they congealed. I was staring at a green dragon, which the ancient Chinese had taken to be the deepest creativity of the universe; and this understanding was breathed into me by my teacher, his face inches from mine.

"So we'll keep the title?" I asked. It wasn't so much a question as my attempt to ground myself in the swirl of what was happening. People had leaned in to hear him speak of the bronze's symbolism, but none of them were able to interpret what he said next. He waited until I looked at him.

"Rarely, if ever, has the universe been presented in such a lyrical manner."

He smiled, his eyes shining.

Epilogue

Our Common Human Destiny

Gentle Reader, I have now finished my narrative of a journey into cosmogenesis during the 1968 to 1983 interval of time. All that remains for me to do is to give a quick summary of the transition into this new life. Conversations with Denise, Bruce, Matt Fox, and Thomas Berry led to our decision to set ourselves up in the San Francisco Bay Area where the journey had begun. Matt and I moved his Institute in Culture and Creation Spirituality from Chicago to Holy Names University in Oakland, California. Bruce switched from the Seattle Mariners to the Oakland Athletics. I became an assistant professor of culture and spirituality, and with a salary of $18,000 a year, a rental on Luella Drive for $650 a month, and public transportation that carried me to a university that supported my work, I felt set for life. Even a century of work would not exhaust the enthusiasm I had for sharing the wisdom Thomas Berry had bequeathed.

My guiding inspiration for teaching cosmogenesis was the magic

Thomas Berry made happen when we sat at his mahogany table surrounded by books. Listening, questioning, and conversing, he led us on a journey from one world to another. I wanted to create courses that reproduced his process. Instead of teaching just the mathematical equations, I would use scientific theories as vehicles for experiencing the cosmological nature of human life. This idea first poked its head up at the University of Puget Sound when I told my students to "ride inside the mathematical symbols." I hardly knew what I meant at the time. That early attempt was my way of stumbling toward a new experience of the universe. It felt then, as it feels today, that we are in the middle of a profound transformation of humanity.

Paleolithic Awakening of Conscious Recall

Our situation in the first half of the twenty-first century has a resonance with an earlier time, when humans first developed the capacity to recall an event from their personal past, employing a new mental capacity now known as the reproductive imagination. Conscious recall is of course commonplace today, but there was a time when this ability was either nonexistent or very rare. For recall to become an aspect of daily life, humans needed to construct a new power of mind. To appreciate this achievement, we need to distinguish the reproductive imagination from other forms of mentality in the animal world. Consider, for instance, the extraordinary mental power exhibited by the red knot sandpiper when it flies nine thousand miles from the southern tip of South America to the Arctic each spring. This is a stupendous feat, but it differs in a qualitative way from the mental capacity I wish to highlight. The difference, both small and stark, shows up at the end of the sandpiper's flight. Once it reaches the Arctic, the sandpiper sets to the important tasks necessary for its survival. It does not retire to a cork-lined room and reflect on the events of its heroic journey. But that is what we do. We developed this new power of mind, reproductive imagination, that enables us to recall events and reflect upon them.

Archeologists have identified at least one of the methodologies our ancestors established for constructing the reproductive imagination. To get a feel for how this took place, begin by imagining we are Cro-Magnon humans entering Rouffignac Cave in Southern France thousands of years ago. Imagine we are crawling on our backs for the journey half a mile down into Earth. In the darkness, we make straight and curvy lines on a rock wall. I say "we make straight and curvy lines" to emphasize the magic that is about to take place. We are not there for the mammalian aims of food or sexual mates. We have made the extraordinary journey on our backs down into the cave for the experience of bringing back the past. Out of the kaleidoscope of our memories, we choose one. A galloping horse. By making dots and dashes on rock, we fix this single memory. Imagine the dark cave with torchlight flickering on the walls. We are astonished. *A horse has appeared.* A few squiggly lines have become a horse, even though no horse was there.

A bizarre event. A new species of apes has taken charcoal, mixed it with saliva, marked up a wall, and suddenly a wild horse springs forth in their common experience. Our ancestors crawled into such caves for sixty thousand years because it was so fulfilling, so entrancing. They did not know it, but they were activating a new capacity of mind. Only because they constructed this power and passed it down to us can I write, thousands of years later, "Imagine we are Cro-Magnon humans entering Rouffignac Cave," and immediately through our imaginations we are inside a cave with torches and rock walls.

The Paleolithics awakened reproductive imagination; we today are awakening time-developmental imagination, the power to experience ourselves as a manifestation of fourteen billion years of cosmological creativity. The Paleolithics had their charcoal and saliva; we have our scientific theories. It might seem outrageous to compare our scientific theories of the universe with their charcoal and saliva, but however different they might be, they perform a similar function. Each is a finger pointing to a hitherto unsuspected world. Paleolithic humans learned to "look through" the charcoal and saliva so that a galloping horse could appear; contemporary humans are learning to look through scientific theories so that deep time, a hidden dimension of matter, can reveal its narrative of primordial plasma, stars, galaxies, a molten Earth. It took early humans tens of thousands of years to establish reproductive imagination as an enduring human capacity. With the acceleration of creativity, perhaps it will only require a century or two for us to establish time-developmental imagination. Like the Paleolithics, we do not have an engineer's outline for how to proceed. But because we are entranced, we have the courage necessary to take our first, tentative steps. Though fitful now, these feelings will become stable in the future. And when that happens, Immanuel Kant's dream will have come true; we will have "evolved into new beings who have learned to see the whole first." When that transformation is completed, seeing the whole first will be as easy as remembering the first time we made love. Throughout the day and in a hundred different ways, we will experience our existence as taking place within the whole, complex, intelligent, living universe.

With my auto-cosmology, I am hoping to entice you into participating in this transformation. That is why, instead of an academic approach, I chose to narrate a couple dozen time-developmental experiences. Moments such as my intimacy with the elegance of the early universe in the Boeing Amphitheater. Or the experience of rising up from the quantum vacuum that came to me walking through the snow of North Tacoma. Or the continuity between gravitational attraction and erotic love, the experience of which liberated me from the habit of Whitehead's fallacy of misplaced concreteness. The two dozen events I presented are my saliva and charcoal on a cave wall. I expressed them in the hope of evoking similar memories of your own. That is what really matters, your experiences, for each of us will awaken to the time-developmental universe in a unique way, and out of these will arise a core set of experiences we share in common. It is in terms of this core set that we will deepen our commonality. I do not know what that core set will be, but I do have a small number of candidates. I offer these as possible help in our planetwide groping for a common experience of who we are and of how we are to live together in this fourteen-billion-year enterprise.

Candidates for Inclusion in a Core Set of Experiences in a Time-Developmental Universe

We humans come forth from the universe the way acorns come forth from the oak tree. Nothing could be more obvious when we reflect on the discoveries of four hundred years of modern science. To experience our derivation from the creative energy of cosmogenesis is to leave behind Descartes's fantasy that the universe is inert, dead matter. The universe created humans. The Milky Way galaxy created humans. Our solar system, with its Sun and planets, created humans. Our Earth, with its rocks and oceans and clouds, created humans. This creativity reveals the nature of the universe. A form of cosmological intelligence drew star dust together and laid down pathways to human intelligence. The words of the previous sentences sound as if they come from a dream, but they are scientific fact. Out of its dark depths, the Virgo supercluster of galaxies invented fish who over a hundred million years constructed the fundamental forms of the vertebrate brains. Though such statements bewilder our modern minds, immersion into time-developmental experience allows us to take in the stunning truth. The process of the universe is primary; we ourselves are constructions of the universe's process.

Rooted in the birth of the universe, humans are cosmic persons drawn toward the future by their fascinations. Each of us can learn to feel ourselves as cosmic persons, can learn to feel our bodies absorbing fourteen billion years of creativity as we awake and arise each morning with quanta of energy coursing through us from the primordial burst. Even in this moment now, carbon and oxygen from stars five billion years ago are assembling themselves as us. Our forms of thought, our layered perceptions of the world, all of these were *invented* by our ancestors and have become us. We are the entire monumental flow of events in the form of a human being.

Astounded by our complex foundations in time, we wonder over the future, over what is coming. We live in a sea of time that perpetually creates itself anew precisely because the universe is unfinished. The ultimate ecstasy for humans is to participate in cosmic development. Our manner of participation is of profound significance because the future of the universe is rooted in our creativity, just as we are rooted in all past achievements of the universe. Though we cannot predict the future's form, we know it will be more astounding than what has already emerged. And it is calling to us. It has always been this way. In the first instants of the universe's existence, the unborn stars were calling to the protons and electrons through gravitational attraction. That is what we feel too. The unborn future calls to us in our experience of fascination. Fascination is how the future speaks. Awakening to, and pursuing, our deepest fascinations is to participate in the joyful difficulty of creating the future.

The universe rests on relationship. The first elementary particles, such as protons, neutrons, and electrons, deepened their relationships and gave birth to a trillion galaxies. No new particles came along. It was done by the original set. I need to say it again. These particles constructed the galaxies by doing one thing: *deepening their relationships.* This mysterious synergy happened again with the emergence of life. Unicellular organisms, each one smaller than the sharp end of a pin, entered into relationships with each other and ended up constructing lions. There is the great mystery. In relationship with another, your deeper identity is ignited. Only by entering into communion with someone outside yourself can you find your true self. Tiny, tiny, individual cells deepened their relationships, united, evolved through time, then flew through the moonless night as great horned owls. In our universe, ultimate creativity rests upon the union of things. Humans took this magic and ran with it. Genetically, we are practically identical with all the other apes, but we found ways to deepen our relationships, which brought forth new capacities to see and to listen. Because of these new capacities, we can now hear the universe tell the story of how it created us, how each of us is a billion-year process.

The universe offers an endless source of energy for our challenge of becoming cosmological beings. The Paleolithics had their gatherings in caves. Later humans had their mosques, churches, temples, synagogues, shrines, and cathedrals. We too will be inventing new ways for drawing into our lives the creative energy of the universe. In a million culturally unique processes, humans will come to experience themselves as modes of the whole creative universe. Looking out at the stars, we will imagine the vastness of a trillion galaxies and will know that we are beholding that which constructed the eyes that are doing the beholding. Our minds will be challenged to make the figure-ground transformation: *the inner is looking at the outer, which has given birth to the inner.* That is the heart of cosmogenesis. Looking at processes that gave birth to our looking. Looking at processes that gave birth both to the carbon of our eyes and the thinking of our minds. As we integrate this revelation into our lives, we ingest an ocean of energy. Though floating in space on a tiny planet, we are also the universe as a whole that has achieved its self-awareness. The same hum fills us again as we leave behind the cramped mind and feel ourselves expand into vast, bottomless energy. The universe out there gave birth to this new awareness. These phrases simply state what mathematical and observational cosmologists have discovered: *when we look out at the night sky, we are looking out at that which is looking.*

BRIAN THOMAS SWIMME is director of the Third Story of the Universe at Human Energy, a nonprofit public-benefit organization, and professor emeritus at the California Institute of Integral Studies in San Francisco, where he taught evolutionary cosmology to graduate students in the Philosophy, Cosmology, and Consciousness program. Mary Evelyn Tucker and Swimme created Emmy Award–winning PBS documentary *Journey of the Universe* and companion book of the same title. With Monica DeRaspe-Bolles and Devin O'Dea, he created the popular YouTube series *The Story of the Noosphere*. His other published works include *The Universe Is a Green Dragon*, *The Universe Story*, written with Thomas Berry, and *The Hidden Heart of the Cosmos*. His three educational series, *Canticle to the Cosmos*, *Earth's Imagination*, and *The Powers of the Universe*, produced by Bruce Bochte, can be streamed at storyoftheuniverse.org.